Lecture Notes in Mathematics

Edited by A. Dold and B. Eckmann
Subseries: Fondazione C.I.M.E., Firenze
Adviser: Roberto Conti

1365

M. Giaquinta (Ed.)

Topics in Calculus of Variations

Lectures given at the 2nd 1987 Session of the Centro Internazionale Matematico Estivo (C.I.M.E.) held at Montecatini Terme, Italy, July 20–28, 1987

Springer-Verlag
Berlin Heidelberg New York London Paris Tokyo

Editor

Mariano Giaquinta
Istituto di Matematica Applicata, Università di Firenze
Via di S. Marta, 3, 50139 Firenze, Italy

Mathematics Subject Classification (1980): 49, 35, 53

ISBN 3-540-50727-2 Springer-Verlag Berlin Heidelberg New York
ISBN 0-387-50727-2 Springer-Verlag New York Berlin Heidelberg

Printed in Germany

Printing and binding: Druckhaus Beltz, Hemsbach/Bergstr.
2146/3140-543210

INTRODUCTION

The international course on "Topics in Calculus of Variations" was held at Montecatini, Italy, July 20-28, 1987, organized by the Fondazione CIME.

These proceedings contain the texts of the lectures presented by L. Caffarelli, J. Moser, L. Nirenberg, R. Schoen, A. Tromba. They also contain the lectures H. Brezis had originally planned and kindly agreed to provide, though he was prevented from coming.

I wish to express my gratitude to the lecturers and to all participants for their contribution to the success of the course. I would also like to express my special thanks to all the authors for undertaking the heavy task of writing the text of their lectures.

Mariano Giaquinta

Firenze, June 1988

TABLE OF CONTENTS

C.I.M.E. Session on "Topics in Calculus of Variations"

List of Participants

L. AMBROSIO, Scuola Normale Superiore, Piazza dei Cavalieri 7, 56100 Pisa

G. ANZELLOTTI, Dipartimento di Matematica, Università di Trento, 38050 Povo (Trento)

A. AROSIO, Via Cocchi 10, 50131 Firenze

M. ATHANASSENAS, Math. Institut, Beringstrasse 4, D-5300 Bonn

P. AVILES, Department of Mathematics, University of California at San Diego, La Jolla, CA 92003, USA

M. BADIALE, Via G. Conti 13, 30014 Cavarzere (VE)

G. BARTUZEL, Institute of Mathematics, Technical University of Warsaw, Plac Jednosci Robotniczej 1, 00661 Warsaw

M.L. BERTOTTI, Dipartimento di Matematica, Università di Trento, 38050 Povo (Trento)

I. BIRINDELLI, Via dei Coronari 82, 00185 Roma

G. BUSONI, Istituto Matematico Università, Viale Morgagni 67/A, 50134 Firenze

L. CAFFARELLI, Institute for Advanced Study, Princeton, N.J. 08540, USA

P. CANNARSA, Dipartimento di Matematica, II Università di Roma, Via O. Raimondo, 00173 Roma

F. CATANESE, Dipartimento di Matematica, Università, Via Buonarroti 2, 56100 Pisa

G. CERAMI, Via G. Sciutti 156, 90144 Palermo

L. FU CHEUNG, Mathematisches Institut der Universitat Bonn, Beringstr.4, 5300 Bonn 1

G. CHITI, Dipartimento di Matematica del Politecnico, Corso Duca degli Abruzzi 24, 10129 Torino

E. COMPARINI, Istituto Matematico Università, Viale Morgagni 67/A, 50134 Firenze

F. DAL FABBRO CROSTA, Dipartimento di Matematica del Politecnico, Piazza L. da Vinci, 32, 20133 Milano

R. DAL PASSO, IAC-CNR, Viale del Policlinico 137, 00161 Roma

G. DEB RAY, St. Xavier's College, 30 Park Street, 700016 Calcutta, India

F. DEL BUONO, Via Eden 15, 52010 Badia Prataglia (Arezzo)

E. DI BENEDETTO, Facoltà di Ingegneria, Università degli Studi di Tor Vergata, Via O. Raimondo, 00173 Roma

B. D'ONOFRIO, Via Veneto 21, 04020 SS. Cosma e Damiano (Latina)

F. DUZAAR, Mathematisches Institut, Universitat Dusseldorf, Universitatsstr. 1, D-4000 Dusseldorf

J. EELLS, I.C.T.P., Box 586, 34100 Miramare, Trieste

H. EGNELL, Department of Mathematics, Thunbergsvagen 3, 752 38 Uppsala, Sweden

J. ESCOBAR, The University of Chicago, Department of Mathematics, 5734 University Av. Chicago, Illinois 60637

A. FASANO, Istituto Matematico Università, Viale Morgagni 67/A, 50134 Firenze

M. FERRARIS, Dipartimento di Matematica, Via Ospedale 72, 090100 Cagliari

G. FIORITO, Via XX Settembre 25, Pal. I, 95017 San Gregorio (Catania)

M. FUCHS, Mathematisches Institut, Universitat Dusseldorf, Universitatsstr. 1, D-4000 Dusseldorf

M. GIAQUINTA, Istituto di Matematica Applicata, Via S. Marta 3, 50139 Firenze

M. GIRARDI, Via Mercalli 21, 00197 Roma

E. GIUSTI, Istituto Matematico Università, Viale Morgagni 67/A, 50134 Firenze

G. GREGORI, Via Venosa 8, 00178 Roma

M. GRUTER, Math. Institut, Universitat Bonn, Beringstr. 6, D-5300 Bonn

F. HELEIN, Ecole Polytechnique, Dept. de Mathematiques, 91128 Palaiseau, France

S. HILDEBRANDT, Drachenfelsstrasse 23, D-5205 St. Augustin 2, W. Germany

L.A. IBORT, Dpto. de Fisica Teorica, Universidad de Zaragoza, 50009 Zaragoza, Spain

P.-A. IVERT, Matematiska Institutionen, Universitetet i Linkoping, S-581 83 Linkoping, Sweden

F. JOSELLIS, Institut f. Reine u. Angewandte Mathematik RWTH Aachen, D-5100 Aachen, W. Germany

M. KOISO, Max-Planck-Institut fur Mathematik, Gottfried Str. 26, D-5300 3, W. Germany

N. KOISO, Max-Planck-Institut fur Mathematik, Gottfried Str. 26, D-5300 Bonn 3, W. Germany

J. KONDERAK, I.C.T.P., P.O. Box 586, Strada Costiera 11, 34100 Miramare, Trieste

P. KUMLIN, Department of Mathematics CTH, S-412 96 Goteborg, Sweden

P. LAURENCE, Courant Institute, 251 Mercer Str., New York, N.Y. 10012, USA

F. LEONETTI, Dipartimento di Matematica pura e applicata, Università de L'Aquila, Via Roma, 67100 L'Aquila

H. LINDBLAD, Matematiska Institutionen, Lunds University, Box 118, 22100 Lund, Sweden

M. LONGINETTI, IAGA-CNR, Via S. Marta 13/A, 50139 Firenze

R. MAGNANINI, Istituto Matematico Università, Viale Morgagni 67/A, 50134 Firenze

G. MANCINI, Dipartimento di Matematica, Piazza di Porta S.Donato 5, 40127 Bologna

P. MARCATI, Dipartimento di Matematica pura e applicata, Università de L'Aquila, Via Roma, 67100 L'Aquila

M. MATZEU, Dipartimento di Matematica, Università "La Sapienza", Città Universitaria, 00185 Roma

A. MAUGERI, Via Etnea 688, 95128 Catania

M. MEIER, Mathematisches Institut, Beringstr. 4, D-5300 Bonn 1

G. MODICA, Istituto di Matematica Applicata, Via S. Marta 3, 50139 Firenze

J. MOSER, ETH-Zentrum, CH-8092 Zurich, Switzerland

M.K.V. MURTHY, Dipartimento di Matematica, Via Buonarroti 2, 56100 Pisa

R. MUSINA, S.I.S.S.A., Strada Costiera, 34014 Grignano, Trieste

L. NIRENBERG, Courant Institute of Mathematical Sciences, 251 Mercer Street, New York, N.Y. 10012, USA

F. NJOKU, S.I.S.S.A., Strada Costiera 11, 34014 Grignano, Trieste

L. NOTARANTONIO, Viale Tirreno 187, 00141 Roma

A. OLVERA, Depto. de Matematicas Aplicadas y Analisis, Facultad de Matematicas, Universidad de Barcelona, Gran Via 585, Barcelona 08071, Spain

F. PACELLA, Via Valsolda 111, 00141 Roma

D. PALLARA, Via Vittorio Veneto 3, 73100 Lecce

G. PALMIERI, Dipartimento di Matematica, Via G. Fortunato, 70125 Bari

B. PELLONI, Via Ronciglione 5, 00191 Roma

M.A. POZIO, Dipartimento di Matematica, II Università di Roma, Via O. Raimondo, 00173 Roma

C. PUCCI, Istituto Matematico Università, Viale Morgagni 67/A, 50134 Firenze

R. PUTTER, Mathematisches Institut, Universitat Bonn, Beringstr. 6, D-5300 Bonn

A. RATTO, Via De Amicis 8/17, 17100 Savona

O. REY, Ecole Polytechnique, Departement de Mathématiques, 91128 Palaiseau, France

R. RICCI, Istituto Matematico Università, Viale Morgagni 67/A, 50134 Firenze

E. ROSSET, Via Isonzo 38, Codroipo (Udine)

R. SCHOEN, Department of Mathematics, Stanford University, Stanford, Ca. 94305, USA

P. SMITH, Max-Planck-Institut fur Mathematik, Gottfried-Claren-Str. 26, D-5300 Bonn 3, W. Germany

K. STEFFEN, Mathematisches Institut, Universitat Dusseldorf, Universitatsstr. 1, D-4000 Dusseldorf

I. STRATIS, Department of Mathematics, University of Athens-Panepistimiopolis, GR 15781 Athens, Greece

D. STROTTMAN, Los Alamos National Laboratory, Los Alamos, New Mexico 87544, USA

M. STRUWE, Mathematik, ETH Zentrum, CH-8092 Zurich, Switzerland

J. SZTAJNIC, Institute of Mathematics, Lodz University, ul. S.Banacha 22, 80-238 Lodz, Poland

G. TALENTI, Istituto Matematico Università, Viale Morgagni 67/A, 50134 Firenze

A. TARSIA, Dipartimento di Matematica, Via Buonarroti 2, 56100 Pisa

N.A. TCHOU, Dipartimento di Matematica, Università "La Sapienza", Città Universitaria, 00815 Roma

B. TERRENI, Piazza Toniolo 10, 56100 Pisa

E. TOMAINI, Via Cimarosa 2, 45100 Rovigo

V.M. TORTORELLI, Scuola Normale Superiore, Piazza dei Cavalieri 7, 56100 Pisa

F. TRICERRI, Istituto Matematico Università, Viale Morgagni 67/A, 50134 Firenze

A. TROMBA, Max-Planck-Institut fur Mathematik, Gottfried-Claren-Str. 26,
D-5300 Bonn 3, W. Germany

L. TUBARO, Dipartimento di Matematica, Università di Trento, 38050 Povo (Trento)

R. TURNER, Department of Mathematics, University of Wisconsin, 480 Lincoln Drive,
Madison, WI 53706, USA

M. UGHI, Dipartimento di Scienze Matematiche, Piazzale Europa 1, 34127 Trieste

G. VALLI, University of Warwick, Coventry, CV4 7AL, (G.B.)

V. VESPRI, Dipartimento di Matematica, II Università di Roma,
Via O. Raimondo, 00173 Roma

G. WEILL, 23 Avenue Marceau, 75016 Paris, France

R. WENTE, Department of Mathematics, University of Toledo, Toledo, Ohio 43606, USA

C.M. WOOD, Department of Pure Mathematics, University of Liverpool, P.O.Box 147,
Liverpool L69 3BX, England

R. YE, Mathematisches Institut der Universitat, Beringstr. 4, 5300 Bonn 1, W. Germany

F. ZIRILLI, Dipartimento di Matematica, Università "La Sapienza",
Città Universitaria, 00185 Roma

S^k – VALUED MAPS WITH SINGULARITIES

Haïm Brézis
Département de Mathématiques, Université Paris 6
4, pl. Jussieu, 75252 Paris Cedex 05
and
Rutgers University, New Brunswick, NJ 08903

The purpose of these notes is to present a survey of some recent results and open problems dealing with the "energy" of S^k – valued maps. The original motivation comes from the theory of liquid crystals; such materials are composed of rod–like molecules with a well defined orientation, except at isolated points (the "defects") which are observed by the physicists. The optic axis φ is a vector of unit length (in $\mathbb{R}^3$) defined in the domain $\Omega \subset \mathbb{R}^3$ (the container of the liquid crystal); so that φ is a map from Ω into S^2. Associated with a configuration φ is a deformation energy which we shall usually take to be

$$E(\varphi) = \int_\Omega |\nabla\varphi|^2 \, dx. \tag{0.1}$$

Physicists consider more general energies, such as,

$$\tilde{E}(\varphi) = \int_\Omega k_1(\mathrm{div}\varphi)^2 + k_2(\varphi\cdot \mathrm{curl}\varphi)^2 + k_3|\varphi\wedge \mathrm{curl}\varphi|^2 + \alpha[\mathrm{tr}(\nabla\varphi)^2-(\mathrm{div}\varphi)^2]dx \tag{0.2}$$

where k_1, k_2, k_3 and α are positive constants. In the special case where $k_1 = k_2 = k_3 = \alpha = 1$, then it is easy to see that $\tilde{E} = E$. While much progress has been achieved for the energy E, little is known so far for $\tilde{E}$. Stable equilibrium configurations correspond to minima of E (or $\tilde{E}$) and therefore it is essential to study the properties of minimizers. For a detailed discussion of the physical background we refer e.g. to [9], [10], [13], [16], [17], [18] and [33]. However we feel that the mathematical questions involved in this field are of great interest for their own sake, an interest which goes much beyond the original motivation. In fact, it is remarkable that progress has been achieved through the joint efforts of experts in Nonlinear Partial Differential Equations, Functional Analysis, Differential Geometry, Geometric Measure Theory, Topology, Numerical Analysis, Graph Theory, etc.

The plan is the following:

I. The problem of prescribed singularities.

I.1. Point singularities in $\mathbb{R}^3$.

I.2. Various generalizations:

1) A domain $\Omega \subset \mathbb{R}^3$ with constant boundary condition.

2) Holes in $\mathbb{R}^3$.

3) An example related to minimal surfaces.

I.3. Some open problems.

II. The problem of free singularities.

II.1. $x/|x|$ is a minimizer.

II.2. The analysis of point singularities.

II.3. Energy estimates for maps which are odd on the boundary.

II.4. The gap phenomenon. Density and nondensity of smooth maps between manifolds. Traces.

II.5. Some open problems.

I. The problem of prescribed singularities.

I.1 Point singularities in $\mathbb{R}^3$.

We start with a simple question (originally raised by J. Ericksen). Let $a_1, a_2, \ldots a_N$ be N points given in $\mathbb{R}^3$ (the desired location of the singularities). Define the class of admissible maps $\mathcal{E}$ to be

$$\mathcal{E} = \{\varphi \in C^1(\mathbb{R}^3 \setminus \bigcup_{i=1}^{N} \{a_i\}; S^2); \int_{\mathbb{R}^3} |\nabla\varphi|^2 < \infty \text{ and } \deg(\varphi, a_i) = d_i \quad \forall i\} \tag{1.1}$$

where the d_i's are given integers, $d_i \in Z$ with $d_i \neq 0$. Here $\deg(\varphi, a_i)$ denotes the Brouwer degree of φ restricted to any small sphere around a_i. [Stable singularities observed by the physicists have always degree ± 1 and the reason why this is so will be given in Section II.2. However it makes sense to formulate the mathematical question with general degrees].

The problem is to study the quantity

$$E = \operatorname*{Inf}_{\varphi \in \mathcal{E}} \int_{\mathbb{R}^3} |\nabla\varphi|^2 dx \tag{1.2}$$

i.e. the least deformation energy needed to produce singularities at a given location with a given degree. Such a question may seem unrealistic because the container Ω is all of $\mathbb{R}^3$ and also because one cannot prescribe physically the location of the singularities. Nevertheless this model problem has interesting features; it has led to the development of new tools which are useful in more realistic questions.

Surprisingly, there is a simple formula for E:

Theorem 1 ([8]). We have

$$E = 8\pi L \tag{1.3}$$

where L is the length of a minimal connection (in a sense to be defined below).

So far, we have made no restriction about the d_i's. However, we must assume that

$$\sum_{i=1}^{N} d_i = 0 \tag{1.4}$$

because of the following:

Lemma 1. $\mathcal{E}$ is nonempty if and only if (1.4) holds.

Sketch of the proof. Suppose first that $\mathcal{E}$ is nonempty and let $\varphi \in \mathcal{E}$. We claim that φ restricted to a large sphere S_R of radius R has degree zero. Intuitively, this is clear because $\int_{\mathbb{R}^3} |\nabla\varphi|^2 < \infty$ implies that, roughly speaking, φ goes to a constant at infinity. More precisely, we recall (see e.g. [36]) that if S is a closed two dimensional surface in $\mathbb{R}^3$ and ψ is a C^1 map from S into S^2 then

$$\deg \psi = \frac{1}{4\pi} \int_S J_\psi \, d\xi \tag{1.5}$$

where J_ψ denotes the Jacobian determinant of ψ. A useful way to write J_ψ is

$$J_\psi = \psi \cdot \psi_x \wedge \psi_y \tag{1.6}$$

where (x,y) are normal coordinates on S. This follows from the fact that $\psi \cdot \psi_x = \psi \cdot \psi_y = 0$, and thus $\psi_x \wedge \psi_y = (J_\psi)\, \psi$. We deduce from (1.5) and (1.6) that

$$|\deg \psi| \leq \frac{1}{8\pi} \int_S |\nabla_T \psi|^2 d\xi$$

where $|\nabla_T\psi|^2 = |\psi_x|^2 + |\psi_y|^2$.

We now return to φ and choose R_1 large enough so that B_{R_1} contains all the singularities a_i. By continuity, $\deg(\varphi_{|S_r})$ is constant for $r > R_1$. We have for any $R_2 > R_1$,

$$\int_{R_1<|x|<R_2} |\nabla\varphi|^2 = \int_{R_1}^{R_2} dr \int_{S_r} |\nabla\varphi|^2 d\xi \geq 8\pi|\deg\varphi_{|S_r}|(R_2-R_1).$$

Letting $R_2 \to \infty$ we see that $\deg \varphi_{|S_r} = 0$ for $r > R_1$.

From the additivity of the degree we conclude that (1.4) holds. The converse is more delicate and follows from an explicit construction sketched in the proof of Theorem 1.

Definition of L, the length of a minimal connection.

It is convenient to start with simple cases:

Case 1: There are only two singularities, a_1 with degree $+1$ and a_2 with degree -1. We shall call this a dipole. Here

$$L = |a_1 - a_2|$$

is the (Euclidean) distance between the two points. Note that it was easy to guess, from dimensional analysis, that E is proportional to a length.

Case 2: All the degrees d_i are equal ± 1. Because of (1.5) there are as many $+$ signs as $-$ signs. We rename the points (a_i) as positive points $p_1, p_2, \ldots, p_k$ and negative points $n_1, n_2, \ldots, n_k$. Then

$$L = \underset{\sigma}{\mathrm{Min}} \sum_{i=1}^{k} |p_i - n_{\sigma(i)}| \tag{1.7}$$

where the minimum is taken over all permutations σ of the integers 1 to k.

Case 3: In the general case, proceed as above except that in the list (p_i, n_i) points are repeated according to their multiplicity $|d_i|$.

Sketch of the proof of Theorem 1. The proof consists of two independent steps:

A) $E \leq 8\pi L$,

B) $E \geq 8\pi L$.

Step A. The main ingredient is the following basic dipole construction:

Lemma 2. Let (a_1, a_2) be a dipole. Given any $\epsilon > 0$ there is a map φ_ϵ which is smooth on $\mathbb{R}^3$, except at (a_1, a_2), such that

$$\deg(\varphi_\epsilon, a_1) = +1, \quad \deg(\varphi_\epsilon, a_2) = -1, \tag{1.8}$$

$$\int |\nabla\varphi_\epsilon|^2 \leq 8\pi|a_1 - a_2| + \epsilon, \tag{1.9}$$

and moreover

φ_ϵ is constant outside an ϵ–neighborhood of the line segment $[a_1, a_2]$.

(1.10)

In fact, given any positive integer d there is a map φ_ϵ which is smooth on $\mathbb{R}^3$, except at (a_1, a_2), such that

$$\deg(\varphi_\epsilon, a_1) = d, \quad \deg(\varphi_\epsilon, a_2) = -1, \tag{1.8'}$$

$$\int |\nabla\varphi_\epsilon|^2 \leq 8\pi|a_1 - a_2|\, d + \epsilon, \tag{1.9'}$$

and (1.10) holds. Such a map is constructed explicitly in [8] (see also [7]). Putting together these basic dipoles over a minimal connection it is easy to prove that $E \leq 8\pi L$. Clearly, this construction also shows that $\mathscr{E}$ is nonempty when (1.4) holds.

Step B. There are two different methods for proving the lower bound $E \geq 8\pi L$. Each one has its own flavor and I will describe both of them.

Proof of (B) via the D–field approach. This is the original method introduced in [8]. To every map φ we associate the vector field D defined as follows

$$D = (\varphi\cdot\varphi_y\wedge\varphi_z,\ \varphi\cdot\varphi_z\wedge\varphi_x,\ \varphi\cdot\varphi_x\wedge\varphi_y) \tag{1.11}$$

where $\varphi_x, \varphi_y, \varphi_z$ denote partial derivatives of φ with respect to x, y, z. A more intrinsic way to define D is to say that D is the pull–back under φ of the canonical 2–form on S^2. The main properties of D are the following:

$$|D| \leq \tfrac{1}{2}\,|\nabla\varphi|^2 \quad \text{on } \mathbb{R}^3 \setminus \bigcup_{i=1}^{N} \{a_i\} \tag{1.12}$$

and

$$\operatorname{div} D = 4\pi \sum_{i=1}^{N} d_i\, \delta_{a_i} \quad \text{in } \mathscr{D}'(\mathbb{R}^3). \tag{1.13}$$

Proof of (1.12). Changing coordinates at a given point we may always assume that

$$\varphi = (0, 0, 1).$$

Since $|\varphi|^2 = 1$ we have $\varphi\cdot\varphi_x = \varphi\cdot\varphi_y = \varphi\cdot\varphi_z = 0$ and thus we may write

$$\varphi_x = (a_1, b_1, 0),\ \varphi_y = (a_2, b_2, 0) \ \text{ and } \ \varphi_3 = (a_3, b_3, 0).$$

We see that

$$D = a \wedge b$$

with $a = (a_1, a_2, a_3)$ and $b = (b_1, b_2, b_3)$.

It follows that

$$|D| \le |a|\,|b| \le \tfrac{1}{2}(|a|^2 + |b|^2) = \tfrac{1}{2}|\nabla\varphi|^2.$$

<u>Proof of (1.13).</u> On $\mathbb{R}^3 \setminus \bigcup_{i=1}^{N} \{a_i\}$ we have

$$\operatorname{div} D = 3\ \varphi_x \cdot \varphi_y \wedge \varphi_z = 0$$

since φ_x, φ_y and φ_z are in the same plane (perpendicular to φ). In view of a celebrated Theorem of L. Schwartz we find

$$\operatorname{div} D = \sum_{\alpha, i} c_\alpha \partial^\alpha \delta_{a_i} \ \text{ in } \ \mathcal{D}'(\mathbb{R}^3).$$

On the other hand, since $D \in L^1(\mathbb{R}^3)$, we must have

$$\operatorname{div} D = \sum_{i=1}^{N} c_i \delta_{a_i} \ \text{ in } \ \mathcal{D}'(\mathbb{R}^3). \tag{1.14}$$

Integrating (1.14) over a small ball B around a_i we see that

$$\int_S D \cdot n = c_i$$

where $S = \partial B$ and n is the outward normal to S. On the other hand, it follows from the definition of D that $D \cdot n = J_\varphi$ where φ is considered as a map *restricted* to S and J_φ denotes its 2×2 Jacobian determinant. Applying (1.5) we find that $c_i = 4\pi \deg(\varphi, a_i)$.

The proof of (B) then proceeds as follows. Let $\zeta\colon \mathbb{R}^3 \to \mathbb{R}$ be any function such that

$$\|\zeta\|_{Lip} = \sup_{x \ne y} \frac{|\zeta(x) - \zeta(y)|}{|x - y|} \le 1,$$

so that $\|\nabla\zeta\|_{L^\infty} \le 1$. We have

$$\int |\nabla\varphi|^2 \ge 2 \int |D| \ge -2 \int D \cdot \nabla\zeta = 2 \sum_{i=1}^{N} 4\pi d_i \zeta(a_i). \tag{1.15}$$

Relabelling the points (a_i) as positive and negative points (p_i, n_i) and taking into account their multiplicity we have

$$\sum_{i=1}^{N} d_i \zeta(a_i) = \sum_{i=1}^{k} \zeta(p_i) - \sum_{i=1}^{k} \zeta(n_i).$$

The conclusion of Step B is a direct consequence of the following:

Lemma 3. Let M be a metric space and let $p_1, p_2, \cdots p_k$ and $n_1, n_2, \cdots n_k$ be 2k points in M. Then

$$\underset{\substack{\zeta : M \to \mathbb{R} \\ \|\zeta\|_{Lip} \le 1}}{\mathrm{Max}} \left\{ \sum_{i=1}^{k} \zeta(p_i) - \sum_{i=1}^{k} \zeta(n_i) \right\} = L$$

where $\|\zeta\|_{Lip} = \sup_{x \neq y} \frac{|\zeta(x) - \zeta(y)|}{d(x,y)}$ and $L = \underset{\sigma}{\mathrm{Min}} \sum_{i=1}^{k} d(p_i, n_{\sigma(i)})$.

A quick proof of Lemma 3 relies on the Kantorovich min–max principle (see [32] or [37]) and the Birkhoff Theorem which asserts that the extreme points of the doubly stochastic matrices are the permutation matrices (see [8] for details). Another self contained proof of Lemma 3 is given in [7].

Proof of (B) via the coarea formula. This new proof discovered by F. Almgren – W. Browder – E. Lieb (see [2]) relies heavily on Federer's coarea formula (see [20], [24] or [41]), which we recall for the convenience of the reader. Suppose φ is a C^1 map from a domain $\Omega \subset \mathbb{R}^n$ into a manifold N of dimension $p \le n$. (Think, for example, of N as being a sphere). The differential of φ, $D\varphi$, is a $(p \times n)$ matrix. Set

$$J_p \varphi = \sqrt{\det(D\varphi \cdot (D\varphi)^t)}$$

where det denotes the determinant of the $p \times p$ matrix $D\varphi \cdot (D\varphi)^t$; $J_p \varphi$ is called the p–Jacobian of φ. We have

$$\int_\Omega J_p \varphi \, dx = \int_N \mathcal{H}^{n-p}(\varphi^{-1}(\xi)) \, d\xi \tag{1.16}$$

where $\mathcal{H}^{n-p}$ is the $(n-p)$–dimensional Hausdorff measure in $\mathbb{R}^n$. In the special case where $N = \mathbb{R}^n$, then $J_n \varphi$ is the usual Jacobian determinant of φ and (1.16) becomes

$$\int_\Omega J_\varphi \, dx = \int_{\varphi(\Omega)} \mathrm{card}(\varphi^{-1}(\xi)) \, d\xi. \tag{1.17}$$

In the case of interest to us we take $\varphi \in \mathcal{E}$, $\Omega = \mathbb{R}^3 \setminus \bigcup_{i=1}^{N} \{a_i\}$ and $N = S^2$. Therefore we find

$$\int_{\mathbb{R}^3} J_2 \varphi \, dx = \int_{S^2} \mathcal{H}^1(\varphi^{-1}(\xi)) \, d\xi. \tag{1.18}$$

First, we claim that

$$J_2 \varphi \le \tfrac{1}{2} |\nabla \varphi|^2. \tag{1.19}$$

With the same notations as in the proof of (1.12) we have

$$J_2\varphi = \sqrt{|a|^2|b|^2-(a\cdot b)^2} = |a\wedge b| = |D| \le \tfrac{1}{2}|\nabla\varphi|^2.$$

Next, we claim that, for a.e. $\xi\epsilon S^2$,

$$\mathcal{H}^1(\varphi^{-1}(\xi)) \ge L. \tag{1.20}$$

This will complete the proof of (B) via the coarea formula.

Proof of (1.20). By Sard's Theorem we know that a.e. $\xi\epsilon S^2$ is a regular value of φ. When ξ is regular value, the Implicit Function Theorm implies that $\varphi^{-1}(\xi)$ is a collection of curves which either connect the points (a_i), or go to infinity, or are closed loops. Here $\mathcal{H}^1(\varphi^{-1}(\xi))$ is the total length of these curves. In view of (1.18), (1.19) and since $\int|\nabla\varphi|^2 < \infty$, the total length is finite and hence there is no curve going to infinity. Furthermore, we shall disregard the closed loops (since they only increase the total length). We are left with a finite collection of curves connecting the points (a_i). Since $\deg(\varphi,a_i) = d_i \neq 0$, at least one curve emanates from each a_i, but there could be more than one. The simplest situation is the case where each positive point p_i is connected by one of the curves to a negative point $n_{\sigma(i)}$, for some permutation σ. Then, clearly $\mathcal{H}^1(\varphi^{-1}(\xi)) \ge L$. Unfortunately, the general situation could be more complicated. For example, a bad configuration would be if we have 4 points p_1, p_2, n_1, n_2 and $\varphi^{-1}(\xi)$ consists only of two curves: one joining p_1 to p_2 and the other n_1 to n_2. We could not conclude, because $|p_1-p_2| + |n_1-n_2|$ might be smaller than L! We shall see that such a configuration is excluded. For this purpose it is convenient to introduce an arrow (i.e. an orientation) on each curve C.

Let x be any point on C and let (e_1, e_2, e_3) be a direct basis with e_1 tangent to C at x. Consider φ restricted to the plane (e_2, e_3) and its (2×2) Jacobian determinant $J_\varphi(x)$. Note that $J_\varphi(x) \neq 0$ since ξ is a regular value. If $J_\varphi(x) > 0$ the orientation of C is given by e_1, and if $J_\varphi(x) < 0$ take the orientation opposite to e_1.

With this convention, and using the properties of the degree, one can see that at every point a_i, one has the basic relation:

$$d_i=\deg(\varphi,a_i)=(\#\text{outgoing arrows})-(\#\text{incoming arrows}). \tag{1.21}$$

For example, an admissible configuration is given by the following figure

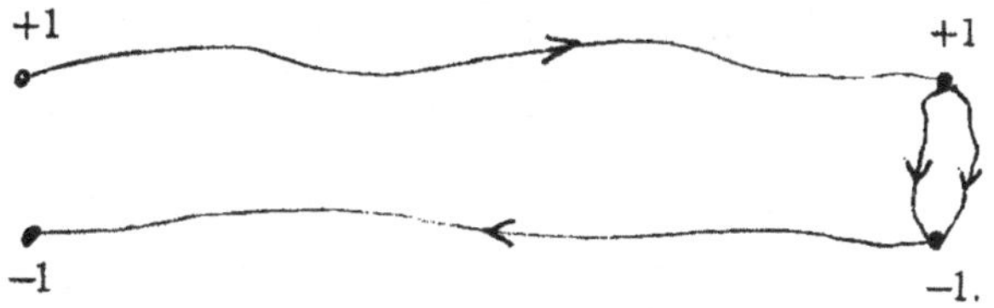

Finally, we claim that any collection of curves running between the points (a_i) and satisfying (1.21) contains a connection connecting each p_i to $n_{\sigma(i)}$ for some permutation σ. Indeed, start with p_1 and take a walk in the direction of the arrow as far as possible. The last point which has been reached is a negative point (because of (1.21)); call it $n_{\sigma(1)}$. Erase this path and note that (1.21) persists. So, we may start again from p_2, etc.

I.2 Various generalizations.

1) A domain $\Omega \subset \mathbb{R}^3$ with constant boundary condition.

Let $\Omega \subset \mathbb{R}^3$ be a smooth bounded domain and let $a_1, a_2, \cdots a_N$ be N points given in Ω. Define the class $\mathcal{E}_1$ to be

$$\mathcal{E}_1 = \{\varphi \in C^1(\bar{\Omega} \setminus \bigcup_{i=1}^{N} \{a_i\}; S^2); \int_\Omega |\nabla \varphi|^2 < \infty, \deg(\varphi, a_i) = d_i \text{ and } \varphi \text{ is constant on } \partial\Omega\}.$$

Again, we assume that $\Sigma d_i = 0$, which is consistent with the condition that φ is constant on $\partial\Omega$. Set

$$E_1 = \operatorname*{Inf}_{\varphi \in \mathcal{E}_1} \int_\Omega |\nabla \varphi|^2.$$

Theorem 2. ([8]). We have

$$E_1 = 8\pi L_1$$

where L_1 is the length of a minimal connection computed using the geodesic distance between points in Ω.

Once more, the proof consists of two steps:

A) $E_1 \leq 8\pi L_1$,

B) $E_1 \geq 8\pi L_1$.

For the proof of (A) the main ingredient is the following extension of Lemma 2:

Lemma 2′. Let C be a simple curve in $\mathbb{R}^3$ with end points (a_1, a_2). Given any $\epsilon > 0$ there is a map φ_ϵ which is smooth on $\mathbb{R}^3$ except at (a_1, a_2), such that

$$\deg(\varphi_\epsilon, a_1) = +1, \quad \deg(\varphi_\epsilon, a_2) = -1,$$

$$\int |\nabla \varphi_\epsilon|^2 \leq 8\pi(\text{length } C) + \epsilon$$

and

$$\varphi_\epsilon \text{ is constant outside an } \epsilon\text{-neighborhood of } C.$$

Proof of (B) via the D–field approach. Consider once more D defined by (1.11). We have, since $D \cdot n = 0$ in $\partial \Omega$, $\int_\Omega |\nabla \varphi|^2 \geq 2 \int_\Omega D \cdot \nabla \zeta = 8\pi \sum_{i=1}^{N} d_i \zeta(a_i)$ for any function $\zeta : \bar{\Omega} \to \mathbb{R}$ such that $\|\nabla \zeta\|_{L^\infty(\Omega)} \leq 1$, i.e. $\|\zeta\|_{\text{Lip}} = \sup\limits_{\substack{x, y \in \Omega \\ x \neq y}} \frac{|\zeta(x) - \zeta(y)|}{d_\Omega(x, y)} \leq 1$ and $d_\Omega(x,y)$ denotes the geodesic distance in Ω. We may then apply Lemma 3 in $M = \Omega$ equipped with the geodesic distance d_Ω.

Proof of (B) via the coarea formula. Once more $\varphi^{-1}(\xi)$ consists of curves in Ω connecting the various points (a_i). These curves do not go to $\partial \Omega$ since φ is constant on $\partial \Omega$ and we may always assume that ξ is different from that constant. We may then proceed as above.

2) Holes in $\mathbb{R}^3$ with prescribed degree.

Assume $H_1, H_2, \cdots H_N$ are disjoint compact sets in $\mathbb{R}^3$. For simplicity, assume each H_i is the closure of a smooth domain. Set

$$\Omega = \mathbb{R}^3 \setminus \left(\bigcup_{i=1}^{N} H_i \right).$$

Consider

$$\mathcal{E}_2 = \{\varphi \in C^1(\bar{\Omega}; S^2); \int_\Omega |\nabla \varphi|^2 < \infty \text{ and } \deg(\varphi, H_i) = d_i \ \ \forall i\}.$$

Here $\deg(\varphi, H_i)$ denotes the degree of φ restricted to ∂H_i. The integers $d_i \in \mathbb{Z}$ are given (possibly zero) with the restriction that $\Sigma d_i = 0$, consistent with $\int_\Omega |\nabla \varphi|^2 < \infty$. Set

$$E_2 = \operatorname*{Inf}_{\varphi \in \mathscr{E}_2} \int_\Omega |\nabla\varphi|^2.$$

Theorem 3 ([8]). We have

$$E_2 = 8\pi L_2$$

where L_2 is the length of a minimal connection connecting the holes and computed using the rule explained below.

Definition of L_2. Given $a, b \in \mathbb{R}^3$ let

$$\delta(a,b) = \begin{cases} 0 & \text{if } a \text{ and } b \text{ belong to the same hole } H_i \\ |a-b| & \text{otherwise,} \end{cases}$$

and

$$D(a,b) = \operatorname{Inf} \sum_{i=1}^{n} \delta(x_{i-1}, x_i)$$

the Infimum being taken over the set of all finite chains $x_0, x_1 \cdots x_n$ joining a to b (i.e. $x_0 = a$ and $x_n = b$). (Note that D is a semi–metric; $D(a,b) = 0$ iff a,b belong to the same hole.) Given two sets A, B, let

$$D(A,B) = \operatorname*{Inf}_{\substack{x \in A \\ y \in B}} D(x,y).$$

Once this reduced distance is defined, throw away the holes of degree zero and relabel the remaining holes as positive holes $P_1, P_2, \cdots P_k$ and negative holes $N_1, N_2, \cdots N_k$ including multiplicity. Set

$$L_2 = \operatorname*{Min}_{\sigma} \sum_{i=1}^{k} D(P_i, N_{\sigma(i)}).$$

Again, the proof consists of two steps:

A) $E_1 \leq 8\pi L_1$

B) $E_2 \geq 8\pi L_2$.

Proof of (A). A minimal connection connecting the holes can be thought of as a finite collection of directed line segments running between pairs of holes and which carry some multiplicity d. On each of these segments one can apply Lemma 2, thereby constructing a map $\varphi_\epsilon \in \mathscr{E}_2$ such that $\int |\nabla\varphi_\epsilon|^2 \leq 8\pi L_2 + \epsilon$.

Proof of (B) via the D–field approach. Let D be defined by (1.11). We have

$$\int_\Omega |\nabla\varphi|^2 \geq 2\int_\Omega |D| \geq 2\int_\Omega D\cdot\nabla\zeta \tag{1.22}$$

for any function $\zeta\colon \mathbb{R}^3 \to \mathbb{R}$ such that $\|\nabla\zeta\|_{L^\infty(\mathbb{R}^3)} \leq 1$. Since $\operatorname{div} D = 0$ in Ω we have

$$\int_\Omega D\cdot\nabla\zeta = \int_{\partial\Omega} (D\cdot\nu)\zeta = \sum_i \int_{\partial H_i} J_\varphi \zeta \tag{1.23}$$

where J_φ denotes the Jacobian determinant of φ restricted to H_i. Assume, in addition, that $\zeta = \zeta(H_i)$ is constant on each H_i, then we have, by (1.5), (1.22) and (1.23)

$$\int_\Omega |\nabla\varphi|^2 \geq 8\pi \sum_{i=1}^N d_i\zeta(H_i) = 8\pi \sum_{i=1}^k \zeta(P_i) - \zeta(N_i)$$

for all functions $\zeta\colon \mathbb{R}^3 \to \mathbb{R}$ such that $\|\nabla\zeta\|_{L^\infty(\mathbb{R}^3)} \leq 1$ which are constant on each hole H_i. This class of functions ζ consists precisely of functions such that $|\zeta(x)-\zeta(y)| \leq D(x,y)$ $\forall x,y\in\mathbb{R}^3$. Therefore we may apply Lemma 3 in $M = \mathbb{R}^3$ equipped with the (semi) metric D and deduce that

$$\int |\nabla\varphi|^2 \geq 8\pi L_2.$$

Proof of (B) via the coarea formula. As above, $\varphi^{-1}(\xi)$ consists of curves in Ω connecting the various holes (H_i). Each curve carries an orientation. For every hole H_i one has the basic relation

$$d_i = \deg(\varphi,H_i) = (\#\text{ outgoing arrows}) - (\#\text{ incoming arrows}).$$

Therefore $\varphi^{-1}(\xi)$ contains a connection connecting each P_i to $N_{\sigma(i)}$ for some permutation σ.

3) An example related to minimal surfaces.

Suppose we change the energy and use instead

$$E'(\varphi) = \int_{\mathbb{R}^3} |\nabla\varphi|\,dx$$

then such an expression has the dimension of an *area*. We are led, by analogy with the previous problem, to pose the following question. Let Γ be an oriented smooth Jordan curve in $\mathbb{R}^3$ and let $\mathcal{E}' = \{\varphi\in C^1(\mathbb{R}^3\setminus\Gamma;S^1);\ \int|\nabla\varphi| < \infty \text{ and } \deg(\varphi,\Gamma) = 1\}$ where $\deg(\varphi,\Gamma) = 1$ means that φ restricted to any small circle which links Γ once has degree 1. Assume $\mathscr{E}'$ is nonempty and set

$$E' = \operatorname*{Inf}_{\varphi\in\mathscr{E}'} \int_{\mathbb{R}^3} |\nabla\varphi|\,dx.$$

The following result, which was conjectured in [8] (and established there for planar curves) has been proved in [2].

Theorem 4. We have

$$E' = 2\pi A$$

where A is the area of an area–minimizing surface spanned by Γ.

Sketch of the proof. The inequality $E' \leq 2\pi A$ is shown by constructing explicitly a map $\varphi\in\mathscr{E}'$ "concentrated" in a neighborhood of a surface spanned by Γ with nearly minimizing area (by analogy with Lemma 2). The reverse inequality $E' \geq 2\pi A$ follows from the coarea formula which takes the form

$$\int_{\mathbb{R}^3} |\nabla\varphi|\,dx = \int_{\mathbb{R}^3} J_1\varphi\,dx = \int_{S^1} \mathscr{H}^2(\varphi^{-1}(\xi))\,d\xi.$$

For a.e. $\xi\in S^1$, $\varphi^{-1}(\xi)$ is a surface spanning Γ and therefore $\mathscr{H}^2(\varphi^{-1}(\xi)) \geq A$.

I.3 Some open problems.

Problem 1. Consider the general energy functional

$$\tilde{E}(\varphi) = \int_{\mathbb{R}^3} k_1(\operatorname{div}\varphi)^2 + k_2(\varphi\cdot\operatorname{curl}\varphi)^2 + k_3|\varphi\wedge\operatorname{curl}\varphi|^2 dx.$$

Is there a simple formula (analogous to (1.3)) for $\operatorname*{Inf}_{\varphi\in\mathscr{E}} \tilde{E}(\varphi)$? The answer is not known even in the dipole case.

Problem 2. Let $\Omega\subset\mathbb{R}^3$ be a smooth bounded domain. Let $g\colon \partial\Omega\to S^2$ be a smooth map of degree zero. Let $a_1, a_2, \cdots a_N$ be N points given in Ω and let

$$\mathscr{E} = \{\varphi\in C^1(\bar{\Omega}\setminus\bigcup_{i=1}^{N}\{a_i\};S^2);\ \int_\Omega |\nabla\varphi|^2 \leq \infty,\ \deg(\varphi,a_i) = d_i\ \forall i \text{ and } \varphi = g \text{ on } \partial\Omega\}$$

where the d_i's are given integers such that $\Sigma d_i = 0$. Set

$$E = \operatorname*{Inf}_{\varphi\in\mathscr{E}} \int |\nabla\varphi|^2$$

and

$$F = 2\,\mathrm{Inf}\,\{\int_\Omega |D|;\ \mathrm{div}\, D = 4\pi \sum_{i=1}^{N} d_i \delta_{a_i} \ \text{ and } \ D\cdot n = J_g \ \text{ on } \ \partial\Omega\}.$$

Is it true (or when is it true) that $E = F$? We always have $E \geq F$; equality is known when g is a constant (see the proof of Theorem 2 via the D–field).

Problem 3. We return to the Example related to minimal surfaces discussed above. The analogue of the vector field D for this problem is the vector field H defined as follows:

$$H = (\varphi \wedge \varphi_x, \varphi \wedge \varphi_y, \varphi \wedge \varphi_z).$$

The main properties of H are

$$|H| = |\nabla\varphi| \quad \text{(this is the counterpart of (1.12))} \tag{1.24}$$

and

$$\mathrm{curl}\, H = 2\pi D_\Gamma \text{(this is the counterpart of (1.13))} \tag{1.25}$$

where D_Γ is a divergence–free measure, supported by Γ, defined by

$$< D_\Gamma, \varphi > = \int_0^1 \varphi(X(t))\cdot X'(t)dt \quad \forall \varphi \epsilon C_0(\mathbb{R}^3;\mathbb{R}^3)$$

and $X(t)$ is any parametrization of Γ. In view of the results above it is natural to ask the question whether

$$\mathop{\mathrm{Inf}}_{\varphi\epsilon\mathscr{E}'} \int_{\mathbb{R}^3} |\nabla\varphi| = 2\pi\, \mathrm{Inf}\{\int_{\mathbb{R}^3} |H|;\ \mathrm{curl}\, H = D_\Gamma\} = 2\pi A\ ?$$

(The answer is positive when Γ is a planar curve see [8].)

II. The problem of free singularities.

Let $\Omega \subset \mathbb{R}^3$ be a smooth bounded domain and let $g\colon \partial\Omega \to S^2$ be a given boundary data. Consider now the problem of minimizing the energy on the class of maps

$$\mathscr{E} = \{\varphi \epsilon H^1(\Omega;S^2);\ \varphi = g \ \text{ on } \ \partial\Omega\}.$$

This class allows, of course, point singularities. Clearly,

$$E = \mathop{\mathrm{Inf}}_{\varphi\epsilon\mathscr{E}} \int |\nabla\varphi|^2 \tag{2.1}$$

is achieved and moreover every minimizer satisfies the equation of harmonic maps

$$-\Delta\varphi = \varphi|\nabla\varphi|^2 \ \text{ on } \ \Omega \tag{2.2}$$

which is a *system* (and not a scalar equation). It is well–known through the work of DeGiorgi [14], Giusti–Miranda [22], Morrey [35], Almgren [1] and others (see a detailed exposition in the book of Giaquinta [21]) that weak solutions of elliptic systems need not be smooth. In our case a result of Schoen–Uhlenbeck [39][40] asserts that every minimizer φ for (2.1) is smooth except at a finite number of points. In contrast with Section I, the number and location of the singularities is not prescribed. If $\deg(g,\partial\Omega) \neq 0$ there is a *topological obstruction* to regularity since g cannot be extended smoothly inside Ω; every map in the class $\mathcal{E}$ must have at least one singularity. If $\deg(g,\partial\Omega) = 0$, there is no topological obstruction to regularity since g can be extended smoothly inside Ω. Nevertheless, we shall see (in Section II.4) that singularities sometimes appear in order to lower the energy of the system.

II.1 $x/|x|$ is a minimizer.

We start with a simple question. Let $\Omega = B^3$ be the unit ball in $\mathbb{R}^3$ and let $g(x) = x$ be the identity map on $\partial\Omega$. The main result is the following:

Theorem 5 ([8]). The map $\varphi(x) = x/|x|$ is a minimizer for (2.1). In fact it is the unique minimizer.

I will present two proofs. The first is the original proof of [8] and is based on the D–field approach. The second proof is a direct and very elegant argument due to F. H. Lin [34]. A third method, using the coarea formula, is described in a more general setting in Section II.3.

Proof of Theorem 5 via the D–field approach. We have

$$|\nabla(\frac{x}{|x|})|^2 = \frac{2}{|x|^2}$$

and therefore

$$\int_\Omega |\nabla(\frac{x}{|x|})|^2 \, dx = 8\pi \int_0^1 dr = 8\pi. \tag{2.3}$$

Therefore it suffices to prove that

$$\int_\Omega |\nabla\varphi|^2\, dx \ge 8\pi, \quad \forall \varphi \epsilon H^1(\Omega;S^2),\ \varphi(x) = x \text{ on } \partial\Omega. \tag{2.4}$$

In fact, we may assume that φ is smooth except at a finite number of points (since we have only to prove (2.4) for minimizers, we may invoke the regularity result of Schoen–Uhlenbeck). Let D be defined by (1.11). We have

$$\int_\Omega |\nabla\varphi|^2 dx \ge 2\int_\Omega |D|\, dx \ge 2\int_\Omega D.\nabla\zeta\, dx \tag{2.5}$$

for any function $\zeta\colon \bar{\Omega} \to \mathbb{R}$ such that $\|\nabla\zeta\|_{L^\infty} \le 1$.

We write

$$\int_\Omega D.\nabla\zeta dx = \int_{\partial\Omega} (D.n)\zeta d\xi - \int_\Omega (\text{div } D)\, \zeta dx \tag{2.6}$$

But $(D.n) = J_g = 1$ on $\partial\Omega$ (since g is the identity on $\partial\Omega$) and $\text{div } D = 4\pi \sum_{i=1}^N d_i \delta_{a_i}$ (the points a_i are the singularities of φ, with the corresponding degrees $d_i \epsilon Z$). Note that

$$\sum_{i=1}^N d_i = 1 \tag{2.7}$$

since g has degree one on $\partial\Omega$.

Applying (2.5) and (2.6) we find

$$\int_\Omega |\nabla\varphi|^2 dx \ge 8\pi \sup_{\substack{\zeta \\ \|\zeta\|_{Lip} \le 1}} \{\int\zeta d\mu - \int\zeta d\nu\}$$

where $d\mu = \frac{1}{4\pi} d\xi$ ($d\xi$ is the surface measure on S^2) and $d\nu = \sum_{i=1}^N d_i \delta_{a_i}$. Since we have no information so far about the location of the points a_i and the integers d_i, except for (2.7), we may as well write

$$\int_\Omega |\nabla\varphi|^2 dx \ge 8\pi \operatorname*{Inf}_{\nu\epsilon\mathcal{A}} \sup_{\substack{\zeta \\ \|\zeta\|_{Lip} \le 1}} \{\int\zeta d\mu - \int\zeta d\nu\} \tag{2.8}$$

where $\mathcal{A}$ denotes the class of measures ν which are finite sums of Dirac masses with integer coefficients (d_i) and $\Sigma d_i = 1$.

Next, we use the following:

Lemma 4. Let M be a compact metric space and let μ be a given probability measure on M. Then

$$\operatorname*{Inf}_{\nu\in\mathscr{A}} \ \sup_{\substack{\zeta:M\to\mathbb{R}\\ \|\zeta\|_{Lip}\le 1}} \{\int_M \zeta d\mu - \zeta d\nu\} = \operatorname*{Min}_{y\in M}\int_M d(x,y)d\mu(x).$$

For the proof of Lemma 4 see [8] or [7] (the proof in [7] relies on a new result in Graph Theory by Hamidoune–Las Vergnas [23]).

We may now complete the proof of Theorem 5 by applying Lemma 4 in $M = \bar{\Omega}$ (with its Euclidean distance). We obtain

$$\int_\Omega |\nabla\varphi|^2 \ge 8\pi \operatorname*{Min}_{y\in\bar{\Omega}} \int_{\partial\Omega} |\xi-y| \frac{d\xi}{4\pi} = 8\pi \tag{2.9}$$

since

$$\int_{\partial\Omega} |\xi-y|d\xi = \frac{1}{2}\int_{\partial\Omega} (|-\xi-y| + |\xi-y|)d\xi \ge \int_{\partial\Omega} |\xi|d\xi = 4\pi.$$

This establishes that $x/|x|$ is a minimizer. For the proof of uniqueness, see [8].

Remark 1. The same argument shows that if Ω is any domain in $\mathbb{R}^3$ and $g\colon \partial\Omega \to S^2$ is a boundary data such that

$$J_g \ge 0 \text{ on } \partial\Omega \text{ and } \deg(g,\partial\Omega) = 1$$

then, for every $\varphi\in H^1(\Omega;S^2)$ such that $\varphi = g$ on $\partial\Omega$, we have

$$\int_\Omega |\nabla\varphi|^2 \ge 2 \operatorname*{Min}_{y\in\bar{\Omega}} \int_{\partial\Omega} d_\Omega(y,\xi)\, J_g(\xi)\, d\xi$$

where d_Ω denotes the geodesic distance in Ω.

Lin's proof of Theorem 5. It relies on the following pointwise inequality which holds a.e. for any $\varphi\in H^1(\Omega;S^2)$:

$$|\nabla\varphi|^2 \ge (\operatorname{div}\varphi)^2 - \operatorname{tr}(\nabla\varphi)^2 \tag{2.10}$$

where $(\nabla\varphi)^2$ is the square of the 3×3 matrix $\nabla\varphi$.

Proof of (2.10). Changing coordinates we may always assume that, at a given point,

$$\varphi = (0,0,1).$$

Set $a_{ij} = \frac{\partial\varphi^i}{\partial x_j}$; since $|\varphi|^2 = 1$ we find $a_{3j} = 0$ for $j = 1,2,3$.

Clearly we have

$$(\mathrm{div}\ \varphi)^2 = (\sum_{i=1}^{2} a_{ii})^2 \leq 2\sum_{i=1}^{2} a_{ii}^2$$

$$\mathrm{tr}(\nabla\varphi)^2 = \sum_{i,j=1}^{3} a_{ij}a_{ji} \geq \sum_{i=1}^{2} a_{ii}^2 - \sum_{i\neq j} a_{ij}^2$$

$$|\nabla\varphi|^2 = \sum_{i=1}^{2} a_{ii}^2 + \sum_{i\neq j} a_{ij}^2$$

which implies (2.10).

On the other hand, a direct computation shows that

$$(\mathrm{div}\ \varphi)^2 - \mathrm{tr}(\nabla\varphi)^2 = \mathrm{div}((\mathrm{div}\ \varphi)\varphi - (\nabla\varphi)\varphi). \tag{2.11}$$

Combining (2.10) and (2.11), and integrating over Ω leads

$$\int_\Omega |\nabla\varphi|^2 dx \geq \int_{\partial\Omega} [(\mathrm{div}\ \varphi)(\varphi.n) - (\nabla\varphi)\varphi.n] d\xi. \tag{2.12}$$

Using the fact that $\varphi(x) = x$ on $\partial\Omega$ it is easy to see that the RHS in (2.12) equals 8π.

Remark 2. Lin's proof extends to any dimension. Let $\Omega = B^n$ be the unit ball in $\mathbb{R}^n$; then $x/|x|$ is a minimizing harmonic map i.e.

$$\int_{B^n} |\nabla\varphi|^2 \geq \int_{B^n} |\nabla(\frac{x}{|x|})|^2 \quad \forall\varphi\epsilon H^1(B^n, S^{n-1}),\ \varphi(x) = x \ \text{on}\ \partial\Omega.$$

In dimension n the counterpart of (2.10) is

$$|\nabla\varphi|^2 \geq \frac{1}{n-2}[(\mathrm{div}\ \varphi)^2 - \mathrm{tr}(\nabla\varphi)^2] \tag{2.10'}$$

while (2.11) remains unchanged. The RHS in (2.12) equals $(n-1)|S^{n-1}|$ and $\int_{B^n} |\nabla(\frac{x}{|x|})|^2 = \frac{(n-1)}{(n-2)} |S^{n-1}|$.

It was already know (see Jager–Kaul [31]) that $x/|x|$ is a minimizer for $n \geq 7$ but this was an open problem for $3 \leq n \leq 6$.

Remark 3. The map $\varphi(x) = x/|x|$ is always a solution of the Euler equation associated to the functional $\tilde{E}$ defined in (0.2), for any choice of the constant k_1, k_2, k_3. A variant of Lin's proofs shows that $x/|x|$ is a minimizer if $k_1 \leq \min(k_2, k_3)$. On the other hand, Hélein [30] has shown that $x/|x|$ is not a minimizer if $8(k_2 - k_1) + k_3 < 0$. The exact range for which $x/|x|$ is a minimizer is not known.

II.2 The analysis of point singularities.

The main result of this Section is a first order expansion near a singularity for a minimizer φ:

Theorem 6 ([8]). Assume $\Omega \subset \mathbb{R}^3$ is any domain and $g: \partial\Omega \to S^2$ is any boundary data. Let φ be minimizer for (2.1). Then all its singularities have degree ± 1. Moreover, for every singularity x_0 there is a rotation R such that

$$\varphi(x) \simeq \pm \frac{R(x-x_0)}{|x-x_0|} \quad \text{as } x \to x_0.$$

Remark 4. Prior to this result, Hardt–Kinderlehrer–Lin [26] had obtained a universal bound for the degree of singularities. Moreover (experimental and) numerical evidence in [11] indicated that this degree is ± 1.

The argument relies on a *blow–up technique* originally introduced in [22] and a careful analysis of the *homogeneous tangent map*. More precisely, assume $x_0 = 0$; as $\epsilon \to 0$ $\varphi(\epsilon x) \to \psi(x)$ (see [38] and [42]) where ψ is a minimizing harmonic map depending only on the direction $x/|x|$, say $\psi(x) = h(x/|x|)$. The punch line of the proof is the following:

Theorem 7 ([8]). Assume $\Omega = B^3$ and $h: \partial\Omega \to S^2$ is any boundary data. Then the homogeneous estension $\psi(x) = h(x/|x|)$ is not a minimizer unless h is a constant or $h = \pm$ Rotation.

The proof of Theorem 7 is quite involved; see [8].

II.3 Energy estimates for maps which are odd on the boundary.

The following result conjectured in [8] is due to Coron–Gulliver [12].

Theorem 8. Assume $\Omega = B^3$ and let $g: \partial\Omega \to S^2$ be an odd boundary condition, i.e. $g(-x) = -g(x)$ on $\partial\Omega$. Then

$$\int_\Omega |\nabla\varphi|^2 \geq 8\pi \quad \forall \varphi \epsilon H^1(\Omega;S^2),\ \varphi(x) = g(x) \text{ on } \partial\Omega. \tag{2.13}$$

In particular, if we choose $g(x) = x$ we obtain (2.4) and deduce that $x/|x|$ is a minimizing harmonic map.

Sketch of the proof. It relies on the coarea formula. Recall that (see Section I.1)

$$\frac{1}{2}\int_\Omega |\nabla\varphi|^2 dx \geq \int_\Omega J_2\varphi dx = \int_{S^2} \mathcal{H}^1(\varphi^{-1}(\xi))d\xi.$$

Here, we cannot assert that for a.e. $\xi \epsilon S^2$, $\mathcal{H}^1(\varphi^{-1}(\xi)) \geq 4\pi$. However, we may write

$$\int_{S^2} \mathcal{H}^1(\varphi^{-1}(\xi))d\xi = \frac{1}{2}\int_{S^2} \mathcal{H}^1(\varphi^{-1}(\xi) \cup \varphi^{-1}(-\xi))d\xi.$$

The conclusion of Theorem 8 follows from:

Claim. For a.e. $\xi \epsilon S^2$, we have

$$\mathcal{H}^1(\varphi^{-1}(\xi) \cup \varphi^{-1}(-\xi)) \geq 2. \tag{2.14}$$

As above, we may always assume that φ is a minimizer. Therefore φ is smooth except at a finite number of singularities (a_i) where φ acts like $\pm$ a rotation. Assuming ξ and $-\xi$ are regular values, the set $C = \varphi^{-1}(\xi) \cup \varphi^{-1}(-\xi)$ consists of a collection of curves running between the points (a_i) and $\partial\Omega$. There may also be closed loops – but we disregard them as in Section I.1. By contrast with the analysis in Section I.1 the singular points (a_i) do not play any special role. In fact, since φ acts like $\pm$ a rotation near the singularities, we find that near each a_i, C looks like a simple smooth curve passing through a_i.

We claim that

$$\varphi^{-1}(\xi) \cap \partial\Omega \text{ consists of k points with k odd.} \tag{2.15}$$

Indeed, since g is odd, we know by Borsuk's Theorem (see e.g. [36]) that $\deg(g,\partial\Omega)$ is odd. On the other hand if we write $(x_i)_{1 \leq i \leq k} = \varphi^{-1}(\xi) \cap \partial\Omega$, then (see e.g. [36])

$$\deg(g,\partial\Omega) = \sum_{i=1}^{k} \operatorname{sign} J_g(x_i)$$

and therefore k must be odd.

Thus the set $C \cap \partial\Omega$ consists of k pairs of antipodal points with k odd.

We split $C \cap \partial\Omega$ as $P \cup Q$ where

$$P = \left\{ x \epsilon \partial \Omega \,\middle|\, \begin{array}{ll} \text{either} & \varphi(x) = \xi \ \text{ and } \ J_g(x) > 0 \\ \text{or} & \varphi(x) = -\xi \ \text{ and } \ J_g(x) < 0 \end{array} \right\}$$

and

$$Q = \left\{ x \epsilon \partial \Omega \,\middle|\, \begin{array}{ll} \text{either} & \varphi(x) = \xi \ \text{ and } \ J_g(x) < 0 \\ \text{or} & \varphi(x) = -\xi \ \text{ and } \ J_g(x) > 0 \end{array} \right\}.$$

Clearly P and Q are antipodal set (i.e. $P = -Q$), each one containing k points. Write

$$P = \{p_1, p_2, \ldots, p_k\} \ \text{ and } \ Q = \{q_1, q_2, \ldots, q_k\},$$

with $q_i = -p_i$.

It is not difficult to check that C consists of k disjoint curves connecting the P points to the Q points (for this purpose, it is convenient to orient the curves in C as in Section I.1). Therefore we find that

$$\text{Length of } C \geq \sum_{i=1}^{k} |p_i - q_{\sigma(i)}| = \sum_{i=1}^{k} |p_i + p_{\sigma(i)}|.$$

For some permutation σ of $\{1,2,\ldots,k\}$. The conclusion follows from

Lemma 5. Assume k is odd and $\{p_1, p_2, \ldots, p_k\}$ are any k points in $\mathbb{R}^n$. Then

$$\sum_{i=1}^{k} |p_i + p_{\sigma(i)}| \geq 2 \min_{1 \leq i \leq k} |p_i|$$

for any permutation σ.

Proof. Suppose first that σ is the permutation $1 \to 2$, $2 \to 3, \ldots, (k-1) \to k$, $k \to 1$. Then we have

$$\begin{aligned} &|p_1 + p_2| + |p_2 + p_3| + \ldots + |p_{k-1} + p_k| + |p_k + p_1| \\ = &|p_1 + p_2| + |-p_2 - p_3| + \ldots + |-p_{k-1} - p_k| + |p_k + p_1| \geq 2|p_1| \end{aligned}$$

(by the triangle inequality).

In the general case we may decompose σ into such elementary cycles. Since k is odd, at least one of them involves k' elements $(1 \leq k' \leq k)$ with k' odd. Then we are reduced to the previous case.

Remark 5. The same argument shows that if $\Omega \subset \mathbb{R}^3$ is any symmetric domain (i.e. $-\Omega = \Omega$) with $0 \epsilon \Omega$ and $g\colon \partial\Omega \to S^2$ is odd, then

$$\int_\Omega |\nabla \varphi|^2 \geq 8\pi \operatorname{dist}(0, \partial\Omega) \quad \forall \varphi \epsilon H^1(\Omega; S^2),\ \varphi(x) = g(x) \ \text{ on } \partial\Omega.$$

Remark 6. The conclusion of Theorem 8 fails if instead of assuming that g is odd one merely assumes that $\deg(g,\partial\Omega)$ is odd. In fact, given any $\epsilon > 0$, it is easy to construct (using Lemma 2) a map $\varphi\epsilon H^1(B^3;S^2)$ such that $\varphi_{|\partial B^3}$ has degree one and $\int_{B^3}|\nabla\varphi|^2 < \epsilon$.

II.4 The gap phenomenon. Density and nondensity of smooth maps between manifolds. Traces.

We start with a very interesting phenomenon discovered by Hardt–Lin [27].

Theorem 9. Let $\Omega = B^3$. There exist smooth boundary data $g\colon \partial\Omega \to S^2$ of degree zero such that

$$\underset{\substack{\varphi\epsilon H^1(\Omega;S^2) \\ \varphi=g \ \text{on}\ \partial\Omega}}{\mathrm{Min}} \int|\nabla\varphi|^2 < \underset{\substack{\varphi\epsilon C^1(\bar\Omega;S^2) \\ \varphi=g \ \text{on}\ \partial\Omega}}{\mathrm{Inf}} \int|\nabla\varphi|^2. \tag{2.16}$$

Note that since g has degree zero there are always maps $\varphi\epsilon C^1(\bar\Omega;S^2)$ such that $\varphi=g$ on $\partial\Omega$.

Sketch of the construction. In fact, given any $\epsilon > 0$ we may choose a g such that, in (2.16), LHS $\simeq \epsilon$ and RHS $\simeq 16\pi$. For this purpose we place two dipoles (a_1,a_2), (b_1,b_2) along the z–axis with $a_1= (0,0,1+\epsilon)$, $a_2= (0,0,1-\epsilon)$, $b_1= (0,0,-1+\epsilon)$ and $b_2= (0,0,-1-\epsilon)$. Using Lemma 2 we obtain a map φ_ϵ which is smooth except at the points $\{a_1,a_2,b_1,b_2\}$ and such that

$$\int|\nabla\varphi_\epsilon|^2 \leq 32\pi\epsilon + 2\epsilon.$$

Define g to be the restriction of φ_ϵ to $\partial\Omega$, so that g is smooth and $\deg(g,\partial\Omega) = 0$. Clearly LHS $\leq 32\pi\epsilon + 2\epsilon$ since we may take φ_ϵ as a testing function. In order to obtain a lower bound for RHS we use the D–field method. Let D be the D–field associated with φ. We have

$$\int_\Omega|\nabla\varphi|^2 \geq 2\int_\Omega|D| \geq 2\int_\Omega D.\nabla\zeta = 2\int_{\partial\Omega}(D.n)\zeta = 2\int_{\partial\Omega}J_g\zeta$$

for every function ζ such that $\|\nabla\zeta\|_{L^\infty} \leq 1$ (here we use the fact that $\varphi\epsilon C^1$, so that $\operatorname{div} D = 0$). Choosing a function ζ such that $\zeta \equiv 0$ near $(0,0,-1)$ and $\zeta \equiv 2-\epsilon$ we obtain, using (1.5),

$$\int_\Omega |\nabla\varphi|^2 \geq 2(2-\epsilon)4\pi.$$

Theorem 9 implies in particular that smooth S^2–valued maps (satisfying a boundary condition) need not be dense in the corresponding Sobolev space.

In the same spirit, it has been pointed out by Schoen–Uhlenbeck [39] that the map $\varphi(x) = x/|x|$ cannot be approximated in the H^1 norm by maps $\varphi_n \epsilon C^1(B^3;S^2)$. Here is a simple proof of this fact using the D–field. Suppose, by contradiction, that such a sequence exists and let (D_n) be the corresponding D–fields. Clearly $D_n \to D$ in $L^1(\Omega;\mathbb{R}^3)$ and $\operatorname{div} D_n = 0$. It follows that $\operatorname{div} D = 0$ in $\mathcal{D}'(\Omega)$. On the other hand we know that $\operatorname{div} D = 4\pi\delta_0$ – a contradiction.

Going to a more general setting we may ask, following Eells–Lemaire [15], whether $C^1(M;N)$ is dense in $W^{1,p}(M;N)$ where M and N are two manifolds (M may have a boundary, but not N). That question has now been completely settled:

Theorem 10. If $p > \dim M$, then $C^1(M;N)$ is dense in $W^{1,p}(M;N)$.

The case where $p > \dim M$ is easy because the Sobolev embedding Theorem implies that $W^{1,p}(M;N) \subset C(M,N)$ and the usual convolution technique can be applied. The case $p = \dim M$ is slightly more delicate and requires a special argument (see [39] and [40]).

Theorem 11. Suppose $p < \dim M$; then $C^1(M;N)$ is dense in $W^{1,p}(M,N)$ if and only if the homotopy group $\pi_{[p]}(N) = 0$.

If $\pi_{[p]}(N) \neq 0$, Bethuel and Zheng [6] have constructed a map $\varphi \epsilon W^{1,p}(M,N)$ which cannot be approximated by smooth maps (their proof uses an earlier result of White [43]). The converse, namely if $\pi_{[p]}(N) = 0$ then smooth maps are dense, is a deep result of Bethuel [4].

As a consequence of Theorem 11 we see that $C^1(B^3,S^2)$ is not dense in $H^1(B^3,S^2)$ because $\pi_2(S^2) \neq 0$, but $C^1(B^3,S^3)$ is dense in $H^1(B^3,S^3)$ because $\pi_2(S^3) = 0$.

When smooth maps are not dense in $W^{1,p}(M,N)$ it is a very interesting problem to study the $W^{1,p}$ closure of smooth maps. A special case has been settled by Bethuel [5]:

Theorem 12. Let $\varphi \epsilon H^1(B^3,S^2)$ and let D be its corresponding D–field. Then there exists a sequence (φ_n) in $C^1(B^3,S^2)$ such that $\varphi_n \to \varphi$ in H^1 if and only if φ satisfies

$$\operatorname{div} D = 0 \text{ in } \mathscr{D}'(B^3). \tag{2.17}$$

Condition (2.17) is clearly necessary but the converse is far from obvious and we refer to [5].

In [4] and [6] there are also interesting results showing that maps in $W^{1,p}(M,N)$ can be approximated by maps which are smooth everywhere except on sets of low dimensions.

Finally, let us mention that Hardt–Lin [28] have studied the trace on $\partial\Omega$ of maps in $W^{1,p}(\Omega;N)$ where $\Omega \subset \mathbb{R}^n$ is a smooth domain. Their main result is:

Theorem 13. Assume $1 < p < n$ and $\pi_i(N) = 0$ for every $0 \le i \le [p]-1$, then any map $g \in W^{1-\frac{1}{p},p}(\partial\Omega;N)$ is a trace of a map $\varphi \in W^{1,p}(M;N)$.

They also present an example of a map $g \in H^{1/2}(\partial B^3;S^1)$ which is not the trace of any map $\varphi \in H^1(B^3;S^1)$.

II.5 Some open problems.

Problem 4. Let φ be a minimizer for the energy $\tilde{E}$ defined by (0.2) in the class $\mathscr{E} = \{\varphi \in H^1(\Omega;S^2);\ \varphi = g \text{ on } \partial\Omega\}$. Is φ smooth except at a finite number of points? [A partial regularity result of Hardt–Kinderlehrer–Lin [25], [26] asserts that φ is smooth except on a set Z with $\mathscr{H}^1(Z) = 0$ and even $\mathscr{H}^\alpha(Z) = 0$ for some $\alpha < 1$.]

Problem 5. Let $\Omega \subset \mathbb{R}^3$ be a bounded smooth domain and let $g\colon \partial\Omega \to S^2$ be a given boundary data. Let $\mathscr{A}$ denote the class of all measures ν of the form $\nu = \Sigma\, d_i \delta_{a_i}$ with $a_i \in \Omega$, $d_i \in Z$ and $\Sigma d_i = \deg(g,\partial\Omega)$, where the points (a_i) are placed arbitrarily in Ω. Does one have

$$\min_{\substack{\varphi \in H^1(\Omega;S^2)\\ \varphi = g \text{ on } \partial\Omega}} \int_\Omega |\nabla\varphi|^2 = \operatorname{Inf}\{\int_\Omega |D|\ ;\ \operatorname{div} D \in \mathscr{A} \text{ and } D.n = \frac{1}{4\pi} J_g \text{ on } \partial\Omega\}?$$

[The answer is positive if $\Omega = B^3$ and $g(x) = x$ on $\partial\Omega$; this follows from the proof of Theorem 5 via the D–field.]

Problem 6. Find a simple proof of Lemma 4. Are there any interesting applications of Lemma 4 in other fields? What happens to $\operatorname*{Inf}_{\nu\in\mathcal{A}} \operatorname*{Sup}_{\zeta}\{\int\zeta d\mu - \int\zeta d\nu\}$ if μ is a signed measure of total mass one (instead of a probability measure)? What happens to $\operatorname*{Inf}_{\nu} \operatorname*{Sup}_{\zeta}\{\ \}$ if $\mathcal{A}$ is replaced by

$$\mathcal{A}_k = \{\nu\,;\,\nu = \Sigma \frac{d_i}{k}\delta_{a_i} \text{ with } d_i \in Z \text{ and } \Sigma d_i = k\}$$

where $k > 1$ is a given integer?

Problem 7. Assume $\Omega = B^n$ $(n \geq 3)$ and $h\colon \partial\Omega \to S^{n-1}$ is a (smooth) boundary data. When is $\psi(x) = h(x/|x|)$ a minimizer for

$$\operatorname*{Min}_{\substack{\rho\in H^1(\Omega;S^{n-1})\\ \varphi=h \text{ on } \partial\Omega}} \int|\nabla\varphi|^2 \,?$$

Problem 8. Let $\Omega \subset \mathbb{R}^3$ be a bounded domain and let $g\colon \partial\Omega \to S^2$ be a boundary data such that $\deg(g,\partial\Omega) = 0$. Let φ be a minimizer for

$$\operatorname*{Min}_{\substack{\varphi\in H^1(\Omega;S^2)\\ \varphi=g \text{ on } \partial\Omega}} \int|\nabla\varphi|^2 .$$

Are there simple conditions on g which guarantee that φ has no singularity? For example, a condition like

$$\int_{\partial\Omega}|\nabla_T g|^2 < 8\pi$$

where ∇_T denotes the tangential gradient. Note that the example described in Theorem 9 satisfies

$$\int_{\partial\Omega}|\nabla_T g|^2 = 16\pi + \epsilon$$

and the corresponding minimizers have singularities. Almgren–Lieb [3] (resp. Hardt–Lin [29]) have obtained estimates on the numbers of singularities of φ in terms of $\|\nabla_T g\|_{L^2}$ (resp. $\|\nabla_T g\|_{L^\infty}$), however there is no control on the size of the constants involved.

Problem 9. Can one prove Theorem 8 using a D–field approach? More precisely, assume $\Omega = B^3$ and let $g\colon \partial\Omega \to S^2$ be an odd boundary condition. Let $D \in L^1(\Omega;\mathbb{R}^3)$ be such that $\operatorname{div} D \in \mathcal{A}$ (defined in Problem 5) and $D.n = \frac{1}{4\pi} J_g$ on $\partial\Omega$. Does on have

$$\int_\Omega |D| \geq 1 \ ?$$

Problem 10. Can one prove Theorem 8 via Lin's device? This boils down to determine whether the RHS of (2.12) is bounded below by 8π for any map φ which is odd on $\partial\Omega$.

Problem 11. Can one extend Theorem 8 to B^n, $n \geq 3$? More precisely, let $\Omega = B^n$ and assume $\varphi \epsilon H^1(\Omega;S^{n-1})$ is odd on $\partial\Omega$. Does one have

$$\int_\Omega |\nabla\varphi|^2 dx \geq \int_\Omega |\nabla(\frac{x}{|x|})|^2 dx = \frac{(n-1)}{(n-2)} |S^{n-1}| \ ?$$

Problem 12. Let $\Omega = B^3$ and let $g\colon \partial\Omega \to S^2$ be a smooth map of degree zero. Is $\mathrm{Inf}\{\int_\Omega |\nabla\varphi|^2 \ ; \ \varphi \epsilon C^1(\bar{\Omega};S^2),\ \varphi = g \text{ on } \partial\Omega\}$ achieved? What is the corresponding problem if $\deg(g,\partial\Omega) \neq 0$?

Problem 13. Let $\Omega \subset \mathbb{R}^3$ be a smooth bounded domain and let $g\colon \partial\Omega \to S^2$ be a (smooth) boundary data. Given $\varphi \epsilon H^1(\Omega;S^2)$ with $\varphi = g$ on $\partial\Omega$ define

$$S(\varphi) = \sup_{\substack{\zeta:\ \Omega \to \mathbb{R} \\ \|\nabla\zeta\|_{L^\infty} \leq 1}} \{\int_\Omega D.\nabla\zeta - \int_{\partial\Omega} J_g \zeta\} \tag{2.18}$$

where D is D–field associated with φ through (1.11).

In the special case where φ is smooth except at isolated singularities (a_i) of degrees (d_i) then $S(\varphi) = 4\pi L$ where L is the length of a minimal connection connecting the singularities (a_i) (computed with the geodesic distance d_Ω). This is a direct consequence of (1.13) and Lemma 3. Of course, if φ has no singularity, then $S(\varphi) = 0$.

Given $\varphi \epsilon H^1(\Omega;S^2)$ with $\varphi = g$ on $\partial\Omega$ it seems of interest to introduce the "modified" energy

$$E^{\#}(\varphi) = \int_\Omega |\nabla\varphi|^2 + 2S(\varphi) \tag{2.19}$$

which takes into account the "interaction" of the singularities.

In view of the example constructed in the proof of Theorem 9 it seems reasonable to ask whether

$$\inf_{\substack{\varphi \epsilon C^1(\bar{\Omega};S^2) \\ \varphi = g \text{ on } \partial\Omega}} \int_\Omega |\nabla\varphi|^2 = \inf_{\substack{\varphi \epsilon H^1(\Omega;S^2) \\ \varphi = g \text{ on } \partial\Omega}} E^{\#}(\varphi) \ ? \tag{2.20}$$

There are some partial results by Bethuel [5] in that direction. Also, is RHS in (2.20) achieved for some φ which is not smooth? Study the properties of $E^{\#}$ with respect to strong and weak convergence in H^1.

Problem 14. Is $C^1(M;N)$ always dense in $W^{1,p}(M;N)$ for the weak topology of $W^{1,p}$? [The answer is positive for $M = B^3$, $N = S^2$ and $p = 2$, see [4].]

Problem 15. Study the density (or nondensity) of $C^1(M;N)$ in $W^{s,p}(M;N)$ where s need not be an integer. Partial results have been obtained by Escobedo [9].

Problem 16. Are the assumptions in Theorem 13 sharp? How does one recognize whether a given map $g \in H^{1/2}(\partial B^3, S^1)$ is the trace of a $\varphi \in H^1(B^3, S^1)$. Same question for general manifold and $1 < p < \infty$.

References

[1] F. Almgren, Existence and regularity almost everywhere of solutions to elliptic variational problems among surfaces of varying topological type and singularity structure, Ann. of Math. 27 (1968), P. 321–391.

[2] F. Almgren – W. Browder – E. Lieb, Co–area, liquid crystals, and minimal surfaces in DD7 – a selection of papers, Springer (1987).

[3] F. Almgren – E. Lieb, Singularities of energy minimizing maps from the ball to the sphere, Bull. Amer. Math. Soc. 17 (1987), p. 304–306 and detailed paper to appear.

[4] F. Bethuel, Approximation dans des espaces de Sobolev entre deux variétés et groupes d'homotopie, C.R. Acad. Sc. Paris (1988) and detailed paper to appear.

[5] F. Bethuel, A characterization of maps in $H^1(B^3,S^2)$, which can be approximated by smooth maps, to appear.

[6] F. Bethuel – X. Zheng, Sur la densité des fonctions régulières entre deux variétés dans les espaces de Sobolev, C.R. Acad. Sc. Paris 303 (1986), p. 447–447 and Density of smooth functions between two manifolds in Sobolev spaces, J. Funct. Anal. (to appear).

[7] H. Brézis, Liquid crystals and energy estimates for S^2–valued maps, in [16].

[8] H. Brézis – J. M. Coron – E. Lieb, Harmonic maps with defects, Comm. Math. Phys. 107 (1986), p. 649–705.

[9] W. Brinkman – P. Cladis, Defects in liquid crystals, Physics Today, May 1982, p. 48–54.

[10] S. Chandrasekhar, Liquid crystals, Cambridge University Press (1977).

[11] R. Cohen – R. Hardt – D. Kinderlehrer – S. Y. Lin – M. Luskin, Minimum energy configurations for liquid crystals: computational results in [16].

[12] J. M. Coron – R. Gulliver, Minimizing p–Harmonic maps into spheres (to appear).

[13] P. De Gennes, The Physics of Liquid Crystals, Clarendon Press, Oxford (1974).

[14] E. De Giorgi, Un esempio di estremali discontinue per un problema variazionale di tipo ellitico, Boll U.M.I. 4 (1968), p. 135–137.

[15] J. Eells – L. Lemaire, A report on harmonic maps, Bull. London Math. Soc. 10 (1978), p 1–68.

[16] J. Ericksen, Equilibrium theory of liquid crystals, in Advances in Liquid Crystals 2, (G.Brown ed.), Acad. Press (1976), p. 233–299.

[17] J. Ericksen, Static theory of point defects in nematic liquid crystals (to appear).

[18] J. Ericksen – D. Kinderlehrer ed., Theory and applications of liquid crystals, IMA Series Vol. 5, Springer (1987).

[19] M. Escobedo, Some remarks on the density of regular mappings in Sobolev classes of S^M–valued functions (to appear).

[20] H. Federer, Geometric measure theory, Springer (1969).

[21] M. Giaquinta, Multiple integrals in the calculus of variations and nonlinear elliptic systems, Princton Univ. Press (1983).

[22] E. Guisti – M. Miranda, Sulla regularita delle soluzioni deboli di una classe di sistemi ellitici quasilineari, Arch. Rat. Mech. Anal. 31 (1968), p. 173–184.

[23] Y. Hamidoune – M. Las Vergnas, Local edge–connecting in regular bipartite graphs (to appear).

[24] R. Hardt, An introduction to geometric measure theory, Lecture Notes, Melbourne Univ. (1979).

[25] R. Hart – D. Kinderlehrer – F. Lin, Existence and partial regularity of static liquid crystal configurations, Comm. Math. Phys. 105 (1986), p. 547–570.

[26] R. Hardt – D. Kinderlehrer – F. Lin, Stable defects of minimizers of constrained variational principles, Ann. I.H.P., Analyse Nonlinéaire (to appear).

[27] R. Hardt – F. Lin, A remark on H^1 mappings, Manuscripta Math. 56 (1986), p. 1–10.

[28] R. Hardt – F. Lin, Mappings minimizing the L^p norm of the gradient, Comm. Pure Appl. Math. 40 (1987), p. 556–588.

[29] R. Hardt – F. Lin, Stability of singularities of minimizing harmonic maps, Math. Annalen (to appear).

[30] F. Hélein, Minima de la fonctionnelle energie libre des cristaux liquides, C.R. Acad. Sc. Paris 305 (1987), p. 565–568.

[31] W. Jager – H. Kaul, Rotationally symmetric harmonic maps from a ball into a sphere and the regularity problem for weak solutions of elliptic systems, J. Reine Angew. Math. 343 (1983), p. 146–161.

[32] L. Kantorovich, On the transfer of masses, Dokl. Akad. Nauk SSSR 37 (1942), p. 227– 229.

[33] M. Kleman, Points, lines and walls, John–Wiley (1983).

[34] F. Lin, Une remarque sur l'application $x/|x|$, C.R. Acad. Sc. Paris (à paraître).

[35] C. Morrey, Partial regularity results for nonlinear elliptic systems, J. Math. Mech. 17 (1968), p. 649–670.

[36] L. Nirenberg, Topics in Nonlinear Functional Analysis, N.Y.U. Lecture Notes, New York (1974).

[37] S. Rachev, The Monge–Kantorovich mass transference problem and its stochastic applications, Theory of Prob. and Applic. 29 (1985), p. 647–676.

[38] R. Schoen – K. Uhlenbeck, A regularity theory for harmonic maps, J. Diff. Geom. 17 (1982), p. 307–335.

[39] R. Schoen – K. Uhlenbeck, Boundary regularity and the Dirichlet problem for harmonic maps, J. Diff. Geom. 18 (1983), p. 253–268.

[40] R. Schoen – K. Uhlenbeck, Approximation theorems for Sobolev mappings (to appear).

[41] L. Simon, Lectures on geometric measure theory, Canberra (1984).

[42] L. Simon, Asymptotics for a class of nonlinear evolution equations with applications to geometric problems, Ann. of Math. 118 (1983), p. 525–571.

[43] B. White, Infina of energy functionals in homotopy classes, J. Diff. Geom. 23 (1986), p. 127–142.

FREE BOUNDARY PROBLEMS
A SURVEY*

Luis A. Caffarelli
I.A.S., Princeton

1. Introduction and examples

In these lectures I will try to describe through three typical elliptic free boundary problems, the basic techniques developed in the last decade for dealing with the regularity of weak solutions and their free boundaries.

A free boundary (or multiphase, or moving boundary) problem can be loosely described as a boundary value problem, for some evolving or stationary physical magnitude or system, for which some of the unknowns or their derivatives change behavior discontinuously through some value of the unknowns.

Simple examples of this behavior are the constrained membrane problem, the flow of one or two flows through porous media or the cavitational flow of one of several ideal fluids.

a) The constrained membrane problems: An elastic membrane in space attached along its boundary, is deformed downwards by a uniform force distribution, but it is not allowed to go below the $\{x_3 = 0\}$ plane.

Variationally, the problem can be described as follows, finding the configuration of minimal energy among admissible configurations, more precisely: given a smooth domain $\Omega \subset R^2$ and a smooth positive function f on $\partial\Omega$ we consider the family of functions v in Ω satisfying

---

*A preliminary version of this research was presented as the Hermann Weyl Lectures, at the Institute for Advances Study, Princeton, in the fall 1985.

a) $v|_{\partial\Omega} = f$ (v is attached along its boundary)

b) $v \geq 0$

c) "v has finite energy"

and among such family of functions we seek the one that minimizes the energy functional

$$E(v) = \int [(\nabla v)^2 + 2v]dx$$

(linearized model) which makes of c),

c) $v \in H^1(\Omega)$, the Hilbert space of L^2 functions with distributional derivatives in L^2.

Here, the minimizer u satisfies (formally) the "Euler" inequalities

a) $u \geq 0$, $\Delta u \leq 1$.

(Since in deducing the Euler equations positive perturbations are always admissible).

b) $u(1 - \Delta u) = 0$.

(Since when $u > 0$ it can be perturbed in both directions and along $u = 0$. Δu cannot be a negative measure).

The free boundary is given implicitly as $\partial\{u > 0\}$, and Δu changes discontinuously from one to zero across it. We will briefly discuss later some basic existence and regularity theory for weak solutions of this problem.

Let us first go to the second example mentioned above, <u>the one or two fluids flow through porous medium</u>:

Here a porous medium $\Omega \subset R^2$ (or R^3) is given, partially saturated by an incompressible fluid in stationary configuration. Different type of data is given along different portions of $\partial\Omega$ for the velocity or the pressure (through impervious areas $\vec{v}\cdot\nu$, the normal component of the flow is zero; where exposed to air the pressure is zero, etc.) and we are interested in reconstructing the saturated region within Ω and the velocity of the flow there.

The simplest model, following Darcy's law and conservation of mass, proposed that on the saturated region

a) $\vec{v} = -\nabla(p + x_3)$.

(v the velocity, p the pressure, x_3 the gravity potential).

b) $0 = \operatorname{div} \vec{v} = \Delta p$

and along the air-liquid interphase the pressure should be zero, (air pressure) and the flow tangential.

That is, in terms of the pressure

a) $p \geq 0$ on Ω, with prescribed Neuman or Dirichlet data on parts of $\partial\Omega$,

b) $p > 0$ defines the unknown flow region, where $\Delta p = 0$,

c) along the air-liquid interface $\partial\{p > 0\}$

$$0 = v \cdot \nu = \nabla p \cdot \nu + \nu \cdot e_3$$

or

$$p_\nu + \nu_3 = 0 .$$

If we have a two dimensional flow of two fluids with densities $\delta_1 < \delta_2$, we can express the problem in terms of a stream function ψ, $(\nabla\psi = (-v_2, v_1)$ for $\vec{v} = (v_1, v_2)$ the velocity vector.

Then, both fluids will be separated by a stream line (say $\psi = 0$), on both flow regions the flow would be incompressible:

a) $\Delta\psi = 0$ in $\{\psi > 0\}$ and $\{\psi < 0\}$.

and the pressure will be continuous along the streamline separating both fluids.

In particular $\nabla p \cdot \delta$, the derivative of p, tangential to the streamline, should be the same from both sides, that is

$$(v \cdot \delta)_{\Omega+} - (v \cdot \delta)_{\Omega-} = (\delta_1 - \delta_2)\delta \cdot e_2 \qquad \text{or}$$

b) $$(\psi_\nu^+ - \psi_\nu^-) = (\delta_1 - \delta_2)\nu \cdot e_1 .$$

along $\psi = 0$.

Here, the speed changes discontinuously when we cross the streamline $\psi = 0$.

In the case of the water-air interphase, the free boundary relations can be described weakly by

$$\Delta p = - \frac{\partial}{\partial x_n} \chi_{p>0} .$$

While in the two fluids flow

$$\Delta\psi = (\delta_1 - \delta_2) \frac{\partial}{\partial x_1} \chi_{\psi>0} .$$

Notice that in the first case, p is restricted to be non-negative, while in the second ψ changes sign across $\psi = 0$.

The third problem can be presented in two dimensions as <u>cavitational flow of one or two perfect fluids</u> or in more dimensions as an equilibrium configuration for <u>heat or electrostatic energy optimization</u>.

Let us present it as the first.

Again we are given a suitable configuration Ω, and Newman or Dirichlet data, for the stream function ψ of two immiscible incompressible and irrotational fluids, separated by the stream line $\psi = 0$.

Then, at each side of such a line, we have $\Delta\psi = 0$, and along the streamline $\psi = 0$, Bernoulli's law (from both sides) demands that

$$\rho_{\pm} | v_{\pm} |^2 + p = C_{\pm}$$

(The pressure should be continuous, and hence $p_+ = p_-$).

We therefore get after a suitable normalization

$$(\psi_\nu^+)^2 - (\psi_\nu^-)^2 = 1 .$$

If $\psi_- \equiv 0$ (or the fluid on $\{\psi < 0\}$ is at rest, we get the one phase problem $\psi \geq 0$, $(\psi_\nu^+)^2 = 1$ along $\partial\{\psi > 0\}$.

Weak solutions of these problems can be obtained by minimizing

$$E(\psi) = \int (\nabla\psi)^2 + \chi_{\psi>0} .$$

2. <u>The local problems we intend to discuss</u> are therefore the following.

<u>Problem I</u>: In the unit ball $B_1 \subset R^n$ we are given a non negative function u, with bounded second derivatives, $(u \in C^{1,1})$.

In the region $\Omega^+ = \{u > 0\}$, u satisfies $\Delta u = 1$. To fix ideas, we will ask that the origin to belong to $\partial\Omega^+$.

The question is then, what can we say about the "surface" $\partial\Omega^+$.

The connection between this problem and example a) is given by a beautiful set of theorems of Lewy and Stampacchia [L-S] and Frehse [F], that show that the unique minimizer of $E(v)$ is in fact a $C^{1,1}$ non negative function satisfying

$$\Delta u = 1 \quad \text{for} \quad u > 0 .$$

Therefore, the situation of problem 1 is a good starting point for the discussion of $\partial\Omega^+$.

The second problem we want to study is

Problem II: In B_1 we are given a continuous, non negative local minimizer of

$$J(u) = \int (\nabla u)^2 + \chi_{u>0} .$$

Again $0 \in \partial\Omega^+(u)$, and we ask information about $\partial\Omega^+$.

Heuristically, from the Hadamard variational formulas one can deduce that if $\partial\Omega^+$ were a smooth surface, we would have $u_\nu = 1$ along it. (See example c)).

Although we ask u to be continuous to talk freely of Ω^+, one can prove that local minimizers are in fact Lipschitz.

Finally, we will discuss the general free boundary problem.

Problem III: In B_1 we are given a weak continuous solution of

a) $\Delta u = 0$ on $\Omega^+(u)$, $[C\Omega^+(u)]^0$

b) $u_\nu^+ = F(u_\nu^-, x, \nu)$ along $\partial\Omega^+ u$.

Describe, again $\partial\Omega^+$. Here F will be smooth in x, ν strictly monotone in u_ν^-, $F(0,x,\nu) > 0$ and $F(+\infty, x, \ ) = +\infty$.

By weak solution we mean one for which we have an existence theorem, namely, whenever $x_0 \in \partial\Omega^+$ and there is a ball tangent to $\partial\Omega^+$ at X_0 from either Ω^+ or $C(\Omega^+)$,

$$u = \alpha \langle X - X_0, \nu\rangle^+ - \beta \langle X - X_0, \nu\rangle^- + o(|X - X_0|)$$

and

$$\alpha = F(\beta, X_0, \nu) .$$

3. A connection with the theory of generalized minimal surfaces:

We chose the three problems above, because their treatment is considerably different and it parallels in a curious way the regularity theory of minimal surfaces.

Let me therefore discuss briefly how this parallelism can be established.

The theory of minimal surfaces as developed by the school of geometric measure theory can be loosely described as follows.

Step 1: Find a family of surfaces, for which its area can be defined in some generalized form.

Make sure that this family is closed under some limiting process and that area is semicontinuous with respect to this process.

Then you can talk about area minimizers among this family.

Example: Boundaries of sets of finite perimeter: a Set Ω has finite perimeter if for any smooth vector field v with $\sup_{x\in\Omega} |v| \le 1$

$$\left|\int_\Omega (\operatorname{div} v)\,dx\right| \le C_0 .$$

The best constant C_0 is called the perimeter of $\partial\Omega$ (notice that heuristically

$$\left|\int_\Omega \operatorname{div} v\,dx\right| = \left|\int_{\partial\Omega} v\cdot\nu\,ds\right| \le |\mathrm{Area}(\partial\Omega)|$$

(For a serious description see [C-D-P]).

Perimeter is then semicontinuous under L^1 convergence of χ_Ω, and the following problem can be solved:

Problem: Let K be a compact set. Among all sets of finite perimeter $\Omega \supset K$, find the one of minimum perimeter.

Solution: Consider a sequence $\Omega_k \supset K$, with

$$\mathrm{per}(\Omega_k) \longrightarrow \inf .$$

and show that $\Omega_k \to \Omega$, with $\mathrm{per}(\Omega) \le \underline{\lim}\ \mathrm{per}\ \Omega_k$.

Of course, the main part of the theory still remains, and that is the second step.

Step 2: Show that $\partial\Omega$ is (except for an unavoidable singular set) a smooth hypersurface satisfying the minimal surface equation (away from K).

To my knowledge, there are mainly three ways of attacking this problem; the three essentially different in nature, exploiting different properties of area minimizing surfaces.

i) the first one exploits the invariance of area minimizing surfaces under rigid motions and expansions and the powerful monotonicity formula. Heuristically, the monotonicity formula says that if S is an area minimizing surface through 0, in R^{n+1}

$$\frac{A(S \cap B\ )}{\rho^n} = \sigma(\rho)$$

is a monotone increasing function of ρ.

One then considers a sequence of expansions $S_k = \{x \in S_k$ if $E_k X \in S\}$ with $E_k \to 0$, shows more compactness of S_k and obtains a limiting "surface" S_0 for which $\sigma(\rho) =$ constant.

A stufy of σ' shows that if $\sigma' \equiv 0$, S_0 must be a cone.

One classifies minimizing cones S_0, studies which are the alternatives to S_0 being a hyperplane, and deduces from there a regularity theorem for S near X_0.

ii) The second approach is in some way a linearization technique, using the fact that in the tangential coordinates the minimal surface equation becomes $\Delta u = 0$, and brings up in play the idea of nonhomogeneous scaling (or blow-up).

It (very loosely) says: suppose that S is a flat enough minimal surface near zero, that is in an appropriate system of coordinates $S \cap B_1 \subset \{|x_{n+1}| < \varepsilon\}$, then in a smaller ball B_δ, in eventually a new system of coordinates, S is still flatter, (that is $S \cap B_\delta \subset \{x'_{n+1}| < < \frac{1}{2}\varepsilon\delta\}$. and $\alpha(e_{n+1}, e'_{n+1}) < C_\varepsilon$. ($\alpha(\ ,\ )$ angle within two vectors).

Proof. Assume there is a sequence of surfaces S_k and $\varepsilon_k \to 0$, for which such an improvement is not possible, expand S_k in the $n+1$ direction, that is $S'_k = \{(x', x_{n+1})/(x', \varepsilon_k x_{n+1}) \in S_k\}$, find (in some sense) a convergent subsequence S'_k, then the limit S'_0 should be the graph of a harmonic function h, contained in the strip $\{|x_{n+1}| \leq 1\}$. But near the origin, such a function is regular, in fact all of its derivatives are bounded by the fact that $|h| < 1$, so the graph of h has a tangent plane to which S_0' stays at least quadratically close in say $B_{1/2}$.

If the compactness we possess on S_k' is strong enough, S_k' stays as close as we want to this new plane.

Pulling back to S_k, we find a contradiction.

iii) Finally, the third technique starts with different regularity assumptions (it suppose S to be the graph of a Lipschitz function) and is analytically the most delicate, but flexible at the same time, using only the minimality of the surface in some very subtle way associated with the maximum principle, that is that minimal surface should not cross each other.

This technique is more precisely the application of the De Giorgi-Moser-Nash theory on Harnack's inequality for solutions of second order elliptic equations with bounded measurable coefficients, in divergence form. Their main theorem says. Let w be an H^1 weak solution of

$$D_i a_{ij} D_j w = 0 \ , \quad \text{in } B_1$$

(a_{ij} bounded measurable, uniformly elliptic).

Then w is Hölder continuous in $B_{1/2}$. It applies to minimal surface theory in the following way: If S is a Lipschitz graph of a function $S = \{x_{n+1} = f(x')\}$, then $D_i f$ is a solution of an equation of the type above, and hence $D_i f$ is Holder or $f \in C^{1,\alpha}$.

In fact let us give a geometric interpretation of this theorem in terms of infinitesimal displacements.

For that purpose let me point out that one of the crucial lemma in De Giorgi's proof is the following.

<u>Lemma</u>: Assume w a solution of the equation above satisfies in B_1

a) $w > 0$,

b) $|\{x : w \geq 1\}| > \varepsilon$

Then $w/B_{1/2} \geq \delta(\varepsilon) > 0$.

How can we view this theorem in the context of infinitesimal translations?

The fact that S is the graph of a Lipschitz function simply means that there is a whole family of translations δ near e_{n+1}, (more precisely a cone $\alpha(\delta, e_{n+1}) \le \gamma_0$) such that if we translate S by $\varepsilon\delta$ ($\varepsilon > 0$, small), then the new surface $S_{\varepsilon\delta}$ will remain above S. ($\gamma_0 = \operatorname{actg} \lambda^{-1}$, λ the Lipschitz constant).

Now, let us call δ a critical translation if δ belongs to the boundary of the cone of admissible translations, that is if $\alpha(\delta, e_{n+1}) = \gamma_0$.

Then it is plausible that for some critical translation δ_0, $S_{\varepsilon\delta_0}$ remains tangent (or much closer than ε) to S. That would be the case if for instance S were a plane with slope exactly λ. But, we contend, for some choice of critical δ, $S_{\varepsilon\delta}$ should get ε away from S.

Heuristically, to see this, let us cover $B_1 \subset R^n$ by a finite number of cones, Γ_k, with aperture $\pi/4$ and look at those points

$$A_k = \{X: \nabla f(X) \in \Gamma_k\}$$

Then, at least for one k, $|A_{k_0}| > \frac{1}{K}$. Assume

$$\Gamma_{k_0} = \{\sigma: \alpha(\sigma, e_1) < \pi/4\}$$

and let δ be the critical translation opposite to e_1,

$$(\delta = -\alpha e_1 + \beta e_n), \ \beta/\alpha = \lambda)$$

Then on A_{k_0}, $S_{\varepsilon\delta}$ should be " $\frac{\beta}{\alpha}$ ε-away " from S.

Now is when the crucial lemma of DeGiorgi takes place. In differential equations language, we are saying:

$$\text{On} |A_{k_0}|, \quad D_{-e_1} f \le 0$$

(or $D_{e_1} f \ge 0$), and on all of B_1, $D_{-e_1} f \le \lambda$, therefore by the lemma above

$$D_{-e_1} f - \lambda \leq - \delta(\frac{1}{K})$$

on $B_{1/2}$.

That is, the translation

$$\delta - -\alpha e_1 + \beta e_{n+1}$$

is not anymore critical in $B_{1/2}$, or S is now "$\varepsilon \frac{\beta}{\alpha}$ -away" from S.

If this argument is carried out carefully, what we prove is that, on a new system of coordinates we have improved the Lipschitz constant of S by a fixed multiple of the previous Lipschitz constant.

If we repeat this argument inductively, we deduce that λ_k the Lipschitz constant in $B_{2^{-k}}$, in an appropriate system of coordinates satisfies $\lambda_k \leq (1-\delta)^k$ that is S is a $C^{1,\alpha}$ surface.

So, to end this section let us briefly discuss the free boundary problems.

Problem I, we will treat with a technique similar to i). Notice that solutions of Problem I are invariant under the scaling

$$U_\lambda = \frac{1}{\lambda^2} U(\lambda X)$$

Therefore, we may hope to have enough compactness so as to "blow up" U by a sequence $\lambda_k \to 0$,

$$U_{\lambda_k} \to U_0 \quad ,$$

a global solution to our problem. We now try to classify U_0 and retrieve from this the information for a free boundary regularity theorem.

Problem II, we could also try by method i), since it is invariant under the family of transformations

$$U_\lambda = \frac{1}{\lambda} U(\lambda X) \quad .$$

Unfortunately, we are not able to classify global solutions. But we do note that $\partial\Omega^+$, for global solutions has a special geometric property, namely $\partial\Omega^+$ has positive mean curvature.

Therefore, we will develop a technique similar to ii). Finally, we will treat the general free boundary problem III and here we develop something akin to the Harnack inequality. That is, we follow this approach:

Let $U_1 \le U_2$ be two solutions in B_1 of the same free boundary problem.

Assume that away from their free boundaries $U_1 < U_2 - \varepsilon$ in a good portion of B_1. Then, we prove, $\partial\Omega^+(U_1)$ "gets ε-away" from $\partial\Omega(U_2)$ in B_1.

(This is like a Harnack inequality for free boundary problems).

Next, in the spirit of our discussion above, we consider a solution U near a "flat" point of the free boundary, that is where for some cone (near e_{n+1} of small translations $U_{\varepsilon\delta} \le U$ (and in particular $\partial\Omega^+(U_{\varepsilon\delta})$ is "above" $\partial\Omega^+(U)$ and by an inductive argument force $\partial\Omega^+(U)$ to become "flatter and flatter" at X_0.

Discussione of Problem I. (for a complete treatment, see [C].)

As we said above, problem I was formulated as follows. On $B_1(0)$ we are given a nonnegative $C^{1,1}$ function U.

On $\Omega^+(U)$, $\Delta U = 1$, and $0 \in \partial\Omega^+(U)$. We want to describe $\partial\Omega^+(U)$ near 0.

We discuss the problem in several steps

1) Regularity and non-degeneracy. As we said above, our regularity hypothesis U has bounded second derivatives is indeed a theorem of Frehse [F]. In fact, it can be seen in [C-K] that the $C^{1,1}$ bound of U in $B_{1/2}$ is a universal one, depending only on the fact that $0 \in \partial\Omega^+$.

Since $0 \in \partial\Omega^+$, it also follows that $\|U\|_{C^{1,1}}$ is preserved under the family of expansions $U_\lambda = \frac{1}{\lambda^2} U(\lambda X)$, and hence one has enough compactness so as to "blow up" the space, that is, consider $\lim_{\lambda_k \to 0} U_{\lambda_k} = U_\infty$.

Clearly U_∞ will be nonnegative, $C^{1,1}$ and $\Delta U_\infty = 1$ whenever $U > 0$. But what assures us that $U_\infty \not\equiv 0$, or in other words, do we have any compactness on $\partial\Omega^+(U_{\lambda_k})$? Can we say that free boundaries converge to free boundaries?

What we need is some non-degeneracy statement about U near $\partial\Omega^+(U)$, that will assure us that free boundaries are stable.

<u>Lemma</u>: <u>If</u> $X_0 \in \overline{(\Omega^+)}$ $\sup_{B_r(X_0)} U \geq \frac{1}{2n} r^2$.

<u>Proof</u>. It is enough to show it for $X_0 \in \Omega^+$.

Consider

$$h(X) := U(X) - \frac{1}{2n} |X - X_0|^2 .$$

Then on $\Omega^+ \cap B_r$, $\Delta h \geq 0$, $h(X_0) > 0$, and $h|_{\partial\Omega^+} \leq 0$.

Then h takes a positive maximum on ∂B_r.

So we now have an <u>optimal regularity and non degeneracy statement</u>:

a) U is $C^{1,1}$ and cannot be better, since ΔU is discontinuous across $\partial\Omega^+$.

b) but, within the constraints that its regularity imposes to U, U "lifts" away from the free boundary as fast as possible.

That is, the first statement allows us to make compactness arguments, the second assures us that the free boundary of the limit is, in some sense, the limit of the free boundaries. (In particular, $\sup_{B_r} U^*_\infty \geq \frac{1}{2n} r^2$ and hence $0 \in \partial\Omega^+(U_\infty)$).

So, now, we would like to classify global solutions. For instance, if $\partial\Omega^+$ was at 0 a differentiable surface, we would have ended up with

$$U_\infty = \frac{1}{2} (X_n^+)^2$$

for an adequate system of coordinates.

But there are unfortunately many other solutions:

$$U = \sum a_i X_i^2$$

with $a_i \geq 0$ and $\sum 2a_i = 1$.

$$\begin{cases} \frac{1}{2n}(r^2 + \frac{1}{n-2}[\frac{2}{r^{n-2}} - n] & \text{for} \quad r \geq 1 \\ 0 \quad \text{otherwise} \end{cases}$$

etc.

The problem is, we lack the powerful monotonicity formula, and in fact, even in two dimensions $\partial\Omega^+$ may have cusp singularities, as shown by Shaeffer [S].

But we are able to partially substitute that with the following property

Lemma: U_∞ is convex.

Or more generally:

Lemma: There is a modulus of continuity $\sigma(r)$ (σ monotone, $\sigma(0^+) = 0$), so that for any i, $D_{ii}U(X) \geq -\sigma(|X|)$.

Proof. By induction, let $M_k \leq \inf_{B_{2^{-k}}} D_{ii}U$. We will find a recurrence relation $M_{k+1} = M_k - f(M_k)$ with $f(r)$ monotone and $f(r) > 0$ for $r > 0$. Obviously, such a relation forces M_k to go to zero.

To find it, let $X_0 \in \Omega^+ \cap B_{2^{-(k+1)}}$, and let $B_r(X_0)$ be the largest ball around X_0 in Ω^+. Since $0 \in \partial\Omega^+$, $r \leq 2^{-(k+1)}$ and $B_r(X_0) \subset B_{2^{-k}}(0)$.

Hence $D_{ii}U \geq -M_k$ on B_r or $D_{ii}U + M_k$ is nonnegative and harmonic in B_r. Then, if somewhere (at Y) in B_r, $D_{ii} + M_k \geq \alpha_0 > 0$, we can apply Harnack's inequality and get

$$D_{ii}U(X_0) + M_k \geq \alpha_0 C(Y,r) .$$

For that purpose, expand B_r to B_1 and U accordingly

$$\text{new } U = \frac{1}{r^2} \text{ old } U(rX)$$

$$\text{new } X_0 = \frac{1}{r} (\text{old } X_0).$$

Let Y be the (new) contact point of $B_1(x_0)$ with $\partial\Omega^+$. There, the direction i (or $-i$) goes inwards B_1 at Y (at worst is tangent).

Let Y_1 be at distance h from Y towards X_0. If, say, $-i$, goes inwards B_1 at Y, we can move in the $-i$ direction from Y_1 a distance $\sim h^{1/2}$ and still stay in B_1. More than that, we can stay at distance $\sim h$ from ∂B_1.

Let Y_2 be the endpoint of such a segment and represent

$$U(Y_2 = U(Y_1) + D_{-i}U(Y_1)\cdot|Y_2 - Y_1| + \underset{\text{(Along the segment)}}{\iint} D_{ii}U(Y)dsdt.$$

But $U(Y_2) \geq 0$, $U(Y_1) \leq h^2$ and $|\nabla U(Y_1)| \leq h$. Since $|Y_2 - Y_1| \sim h^{1/2}$, we get

$$\underset{\text{(Along the segment)}}{\iint} D_{ii}U\, dsdt \geq -C\, h^{3/2}\ .$$

But $|Y_2 - y_1| \sim h^{1/2}$ therefore

$$\underset{\text{(Along segment)}}{\sup} D_{ii}\, U \geq C\, h^{1/2}\ .$$

We now want to chose h to have some gain on the estimate

$$D_{ii}U + M_k \geq 0.$$

That is, we want

$$-C\, h^{1/2} \geq \frac{-M_k}{2}\ .$$

So we can say, there exists a Y_3 in this segment, so that

$$D_{ii}U(Y_3) + M_k \geq \frac{M_k}{2}\ .$$

This forces $d(Y_3,\partial B_1) \sim M_k^2$.

Therefore Harnack's inequality gives

$$D_{ii}U(X_0) + M_k \geq -\frac{M_k}{2}(M_k)^{2n}$$

or

$$-M_{k+1} \geq -M_k - CM_k^{2n+1} ,$$

and the proof is complete.

At this point we know that global solutions (and hence their free boundaries) are convex.

This is "half" a regularity statement. To get the "other half" we have to contrapose once more the regularity of U, against the convexity of $\partial\Omega^+$. In other words, we want to show that $\partial\Omega^+$ cannot be asymptotically a cone.

For that purpose, blow up once more U_0 around 0, that is look at

$$(U_0)_0 = \lim_{\lambda_k \to 0} (U_0)_{\lambda_k} .$$

Then, $\Lambda = \{(U_0)_0 = 0\}$ is a point or a convex cone. If it is a point, or a convex cone with empty interior, $\Delta(U_0)_0 = 1$ across it, and since $(U_0)_0 \geq 0$, it is a quadratic polynomial.

In this case, there is nothing we can say about $\partial\Omega^+(U)$ near 0. If Λ has non empty interior and δ is a vector in its interior, then Λ is a half space, and $U = \frac{1}{2}(X_M)^2$. (We have finally reached the problem of classifying global "cone" solutions of our problem, and we will replace the monotonicity formula, with the convexity and regularity of U.) Indeed, let δ be a vector in Λ^0.

Then $D_\delta U$ vanishes identically in $\partial\Lambda = \partial\Omega^+$ and $D_\delta U$ is negative (since $D_{\delta\delta} > 0$), Lipschitz and harmonic in Ω^+. This is not possible at the vertex of Λ, unless Λ is a half space. (If Λ has two different planes of support, the two dimensional barrier $h(r,\theta) = -\phi^{1-\varepsilon}$ $\cos(1-\varepsilon)\theta$ is an upper barrier for U in the proper system of coordinates). What is the theorem that this global classification induces in the original $\partial\Omega^+(U)$? It is a dicotomy statement: If $\mathcal{C}(\Omega^+)$ is very thin near 0, there is nothing we can say, but if not, $\partial\Omega^+$ is a smooth surface near 0.

More precisely: We measure the thickness of $\mathcal{C}\Omega^+$ through the normalized minimum diameter:

$$m.d.(r) = m.d.(C\Omega \cap B_r) = \text{infimum width among all strips containing } C\Omega^+ \cap B_r$$

$$n.m.d.(r) = \frac{m.d.(r)}{r}$$

<u>Theorem</u>: There exists a modulus of continuity $\sigma(r), \sigma(0^+) = 0)$ such that

<u>either</u> $$n.m.d.(r) \leq \sigma(r) \quad \forall\ r.$$

<u>Or</u>: If for at least one r

$$n.m.d.(r_0) \geq \sigma(r_0)$$

then $\partial\Omega^+$ is a C^1 surface (the C^1-graph of a function) in a neighborhood of zero.

If then follows from a theorem of Kinderlehrer and Nirenberg [K-N], that $\partial\Omega^+$ is analytic.

<u>Problem II</u>: (For details see [A-C]. In B_1, we have now a nonnegative (continuous) local minimizer U, of the functional

$$J(U) = \int (\nabla U)^2 + X_{U>0} dx ,$$

that is $u \in (H')^+$ and if $v \in (H')^+$, $u = v$ near ∂B_1, $U(u) \leq J(v)$.

As before, to fix ideas we assume $0 \in \partial\Omega^+$. To guide us through our heuristic reasoning, let us point out that we expect u to satisfy $u_\nu = 1$ along $\partial\Omega$.

Indeed, the Hadamard variational formula (the equivalent of Euler equations along $\partial\Omega^+$) say that if we perturb $\partial\Omega^+$ (outwards) an amount δv, the Dirichlet integral decreases by about $(u_\nu^+)^2 \delta v + o(\delta v)$.

Since the terms in the functional must balance each other we get $-(u_\nu^+)^2 \delta v + \delta v = 0$ or $u_\nu^+ = 1$.

<u>Step one</u>: Optimal regularity and non degeneracy: The first observation we can make is the fact u cannot be globally better than Lipschitz since $|\nabla u|$ is discontinuous across $\partial\Omega^+$.

So Lipschitz is the optimal regularity we could hope for. On the other hand let me convice you that on $B_{1/2}$, u should satisfy

$$u(x) \leq Cd(x,\partial\Omega^{+}) = Cd.$$

Indeed, u is positive (and hence harmonic) in $B_d(x)$. If $u(x) \geq Md$ then, by Harnack's inequality $u \geq CMd$ in $B_{d/2}(x)$. Let h be the harmonic function in $B_d \setminus B_{d/2}$ satisfying

$$h/_{\partial B_d} = 0$$

$$h/_{\partial B_{d/2}} = CMd$$

Then $u \geq h$ in $B_d \setminus B_{d/2}$. Look now at $Y \in \partial B_d \cap \partial\Omega^{+}$. There, $u = h = 0$ and hence $u_\nu \geq h_\nu \geq CM > 1$ for M large. A contradiction to the condition $u_\nu = 1$.

Since this reasoning is not yet allowed to us, we exchange it by an integral one. But before, let me point out that problem II is invariant under the family of transformations $u_\lambda = \frac{1}{\lambda} u(\lambda x)$. (That is, if u_λ is a local minimizer of J, u is also such).

Therefore we can normalize the situation to $d = 1$ (note that precisely u_λ preserves Lipschitz norms).

So we argue as before: If $u > M$ at X and $d(X,\partial\Omega^{+}) = 1$ then $u > CM$ on $B_{1/2}(X)$ (by Harnack).

Let $Y \in \partial B_1(X) \cap \partial\Omega^{+}$ and g the harmonic function in $B_1(Y)$ satisfying

$$g/_{\partial B_1(Y)} = u .$$

We will replace u by g, in $B_1(Y)$ and compare $J(u)$ with $J(g)$.

Note that $\displaystyle\fint_{\partial B_1(Y)} g \sim M$ (singe $g \sim M$ in $B_{1/2}(X)$). Hence $g(Y)$ and $g/B_{1/2}(X) \sim M$. Using the h above as a barrier

$$h/_{\partial B_1(Y)} = 0,\ h/_{\partial B_{1/2}(Y)} = M ,$$

h harmonic in between, we get

$$g \geq CM\, d(X, \partial B_1(Y)).$$

We are ready to compare $J(u)$ with $J(g)$. On one hand $|\chi_{u>0}| \leq |\chi_{g>0}| = |B_1|$. Ore more precisely $|\chi_{g>0}| - |\chi_{u>0}| = |\chi_{u=0}|$. On the other, since g is harmonic, and $u - g = 0$ on ∂B_1 we have

$$\int |\nabla u|^2 - \int |\nabla g|^2 = \int |\nabla u - g|^2 .$$

Since

$$\{u = 0\} \subset \{|u - g| > Md\}$$

$$|\{u = 0\}| \leq \frac{1}{M^2} \int |\nabla (u - g)|^2 ,$$

this makes $J(u) > J(g)$ a contradiction unless $u \equiv g$.

But this is in turn, a contradiction since $u(0) = C$. Therefore, $u(x) \leq Cd(x, \partial\Omega^+)$ and hence $|\nabla u| \leq C$ (that is u is Lipschitz).

That u is non degenerate is much simpler. In fact, u non degenerate in this case means that u must grow linearly away from $\partial\Omega^+$.

Indeed, let us show that if (after normalization) $d(x, \partial\Omega^+) = 1$, than $u(x) \geq C$.

Let ϕ be a C_∞ function satisfying $\phi \equiv 0$ on $B_{1/4}(x)$ $\phi \equiv 1$ outside $B_{1/2}(x)$.

If $u(x) = \varepsilon$, then $u/_{B_{1/2}} \leq C\varepsilon$ and hence $w = \min(u, C\varepsilon\phi)$ is well defined and an admissible function. We compare $J(w)$ with $J(u)$

$$\int |\nabla w|^2 \leq \int |\nabla u|^2 + (C\varepsilon)^2 \cdot \int |\nabla \phi|^2 |\chi_{w=0}| = |\chi_{u=0}| - |B_{1/4}| .$$

For ε small this is a contradiction.

This completes the optimal regularity and non degeneracy part of the theory.

This allows us to look for blow up sequences and global solutions u_0 to our problem. It also implies very strong geometrical restrictions for $\partial\Omega^+$. $\partial\Omega^+$ has finite H^{n-1}-Hausdorff measure, (furthermore $H^{n-1}(\partial\Omega^+ \cap B_r) \leq r^{n-1}$ and $\Delta u = g(x)dH^{n-1}$ with $0 < \varepsilon \leq g(x) \leq \frac{1}{\varepsilon}$.

Furthermore, it is not hard to prove that the only half space solutions are x_n^+, and hence $g(x) = 1$.

Since heuristically $\Delta u = u_\nu^+ \, dA$, the free boundary condition is satisfied by now in a very reasonable sense. This also implies, by a blow up argument that

$$\overline{\lim_{X \to X_0 \in \partial\Omega^+}} (\nabla u(x)) = 1.$$

Indeed, let X_k be a sequence for which $\overline{\lim}$ is realized, Y_k the closest point in $\partial\Omega^+$ and normalize so that

$$Y_k = 0, \quad X_k \equiv e_n.$$

Then, for the limiting function u, $|\nabla u|$ has an interior maximum at e_n. Being harmonic u is linear in the connected component of $\Omega^+(u)$ containing e_n, with slope $\overline{\lim}$.

But then $u/_{R_n^+} = x_n^+$ and $\lim = 1$.

<u>Corollary 1</u>: <u>The blow up global solutions</u> u_0 <u>satisfy</u> $|\nabla u_0| \leq 1$.

This has a very important geometric consequence for $\partial\Omega^+$.

<u>Corollary 2</u>: $\partial\Omega^+$ <u>is a generalized surface of positive (outwards) means curvature</u>

What do we mean by this last statement? We mean that if S' is a surface that coincides with $\partial\Omega^+$ outside a small compact set K and S' is inside Ω^+ in K (at one side of $\partial\Omega^+$) then $H^{n-1}(S') \geq$ $\geq H^{n-1}(\partial\Omega^+)$. This can be formally expressed, for instance by saying if $\bar{\Omega} \cap CB_M = \Omega^+ \cap CB_M$ and $\bar{\Omega} \subset \Omega^+$.

$$\operatorname{Per}(\bar{\Omega})|_{B_M} \geq \operatorname{Per}(\Omega^+)|_{B_M}) \ .$$

The reason is very simple, let us apply the divergence theorem to ∇u in the domain $\Omega^+ \subset \bar{\Omega}$.

$$0 = \int \Delta u = \int_{\partial(\Omega^+ \setminus \bar{\Omega})} u_\nu \, dA = \int_{(\partial\Omega^+ \cap B_M)} u_\nu + \int_{\partial\bar{\Omega} \cap B_M} u_\nu$$

(with the proper orientation for u_ν).

But $\int_{\partial\Omega^+ \cap B_M} u_\nu = \text{Per } \partial\Omega^+|_{B_M}$ since $u_\nu = g(x) \equiv 1$, and $\int_{\partial\bar{\Omega} \cap B_M} u_\nu \le \text{Per } \partial\bar{\Omega}|_{B_M}$ since $|\nabla u| \le 1$ everywhere. As one can see, this is as in problem I, a one sided geometrical estimate on $\partial\Omega^+$, but unfortunately much weaker than convexity.

In order to exploit it, we will have to resort, then, to a method much in the style of ii). That is, we plan to "linearize" it to a subharmonic function, and use the regularity of u to obtain the other side estimate.

The linearization process is a delicate matter and we will avoid a detailed discussion here.

The main theorem can be stated as follows:

<u>Theorem</u>: There exists a critical value ε_0, and μ_0, λ_0 with $0 < \mu_0 < 1$, $0 < \lambda_0 < 1$, so that if $\partial\Omega^+ \cap B_1$ is $\varepsilon < \varepsilon_0$ flat, that is, it is contained in $\{|x_n| < \varepsilon\}$, then for some $\rho > \mu_0$ $\partial\Omega^+ \cap B_\rho$ is $\lambda_0\varepsilon_0$ flat, that is, it is contained in $\{|x_n'| < \rho\lambda_0\varepsilon_0\}$ (in an adequate system of coordinates, with $\alpha(e_n, e_n') < C\varepsilon$.

It is not hard to see that such a theorem implies $\partial\Omega^+$ is a Holder differentiable function in a neighborhood of a "flat point" (in particular a point of H^{n-1}-differentiability).

The lines of the proof are as follows: We consider a sequence of flatter and flatter free boundaries $S_{\varepsilon_k} \subset \{|x_n| < \varepsilon_k\}$ (with $\varepsilon_k \to 0$), and we expand them in the x_n direction:

$$\bar{S}_{\varepsilon_k} = \{(x', x_n)/(x', \varepsilon_k x_n) \in S_{\varepsilon_k}$$

The $\bar{S}_{\varepsilon_k}$ converge to a subharmonic function $x_n = h(x')$. Since we want to balance subharmonicity of $h(x')$ with the regularity of u, we simultaneously keep track of

$$v_k = \frac{1}{\varepsilon_k}(u_k + x_n),$$

below S_{ε_k}.

We prove that $v_k \to v$ (harmonic), on $B_1 \cap \{x_n < 0\}$, that $v|_{x_n=0}(x') = h(x')$ and that $v(0,-\lambda) \le 0$ (since $|\nabla u_k| < 1$ and $v_k(0) = 0$.

That is, through our linearization we have kept track of the limit of the S_ε and the deviation of u_k from the half plane solution.

We now balance the subharmonicity of $h(x)$ with the fact that $v_\nu \le 0$ at 0. In terms of Poisson integral representation we get:

$$\frac{1}{\delta}\int \frac{h(x')\delta}{(\delta^2 + (x')^2)^{n/2}}\, dx' \le C$$

or in polar coordinates,

$$\int \frac{h(\rho,\sigma)\ \rho^{n-2}}{(\delta^2 + \rho^2)^{n/2}}\, d\rho d\sigma .$$

Since h is subharmonic $\int h(\rho,\sigma)d\sigma$ is positive, increasing from $h(0) = 0$. Therefore

$$\int \frac{1}{\rho^2}\left[\int h(\rho,\sigma)d\sigma\right]d\rho .$$

Strongly indicating that

$$\fint_{\partial B_\rho} h(x) = \int h(\rho,\sigma)d\sigma$$

should go to zero faster than ρ. Some real analysis work then completes the proof of our theorem.

Problem III: The discussion of problem III is probably the most intricate of all, and we will only dwell on the main new ideas of this approach.

The approach consists of three parts: a) An existence theorem for weak solutions, b) showing that if the free boundary is flat near a point χ_0, then it is the graph of a Lipschitz function and then, that this graph is really a $C^{1,\alpha}$ surface.

The existence theory, relies on properly defining the notion of super and subsolutions to our problem and showing that given a supersolution as upper barrier and a subsolution as lower, then the least of all supersolutions above the lower barrier is a weak solution.

The definition of supersolution is as follows: if $u(x_0) = 0$ and for some ν,

$$u(x) \geq \alpha\langle x - x_{0-},\nu\rangle^+ + o(|x - x_0|), \text{ for } \langle x - x_0,\nu\rangle \geq 0,$$

then for $\langle x - x_0,\nu\rangle \leq 0$

$$u(x) \leq =\beta\langle x - x_0,\nu\rangle^- + o(|x - x_0|)$$

The question is of course, what kind of compactness do we expect in order to show that the limiting function (the inf w, w a supersolution) is in any sense a weak solution.

The necessary compactness is given by the following monotonicity formula (see [A-C-F]).

<u>Theorem</u>: Let u_1, u_2 be two non negative subharmonic functions in disjoint domains $\Omega_1 \cap B_r$, $\Omega_2 \cap B_r$ and vanishing continuously at its boundaries $(\partial\Omega_i)$. Then

$$\sigma(r) = \frac{\iint_{\Omega_1 \cap B_r} |\nabla u_1|^2 \rho d\rho d\sigma \iint_{\Omega_2 \cap B_r} |\nabla u_2|^2 \rho d\rho d\sigma}{r^4}$$

is a monotone function of r.

(Note that if u_1, u_2 are linear functions $\sigma(r)$ is constant). As an application, let us show that if w is a supersolution w^+ is Lipschitz.

Indeed let $d(x_0,\partial\Omega^+(u)) = h$. Then we show that $w(x_0) \leq Ch$. If $w(x_0) > Mh$ then $w(x) \geq CMh$ on $B_{h/2}$ (by Harnack). By using a radial barrier as in problem II, we get

$$w(x) > CMd(x, \partial B_h)$$

Let $Y_0 \subset \partial B_h \cap \partial\Omega^+$, then, if ν is the inner radial direction $(\nu = \frac{X_0 - Y_0}{|X_0 - Y_0|})$, we have

$$w \geq CM\langle Y - Y_0, \nu\rangle^+ \quad \text{on} \quad B_h.$$

By definition of supersolution,

$$w \leq -\beta\langle Y - Y_0, \nu\rangle^- \quad \text{on} \quad B_h$$

for any β such that $C(\beta, Y_0, \nu) \leq CM$. It follows that $\sigma(0^+) \geq CM^2\beta^2$. But for M large, β becomes large, and $\sigma(0^+) \leq \sigma(1) \leq \|u\|_{L^2} \leq C_0$ a contradiction.

We will assume then, an existence theorem of the following type.

Setting of problem III: We are given in B_1 $u_\nu^+ = C(u_\nu^-, X, \nu)$ in the following sense and with the following properties

i) u is Lipschitz and u^+ is non degenerate (that is $u^+(X_0) \geq$ $\geq d(X_0, \partial\Omega^+)$, $\Delta u = 0$ on Ω^+ and $[\mathcal{C}(\Omega^+)]^0$.

ii) In particular $\partial\Omega^+$ has locally finite Hausdorff measure and

$$H^{n-1}(\partial\Omega^+, B_r) \leq Cr^{n-1}$$

iii) If $X_0 \in \partial\Omega^+$ and $B_\varepsilon(Y_0) \subset \Omega^+$ (or $\mathcal{C}\Omega^+$) is tangent to $\partial\Omega^+$ at X_0 then

$$u(X) = \alpha\langle X - X_0, \nu\rangle^+ - \beta\langle X - X_0, \nu\rangle^- + o(|X - X_0|)$$

with ν the inner normal to Ω^+ defined by B_ε, and $\alpha = C(\beta, X_0, \nu)$.

We will discuss the following theorem:

w) If $\partial\Omega^+$ is flat enough in B_1 (that is

$$B_1 \cap \{x_n > -\varepsilon\} \supset \Omega^+ \cap B_1 \supset \{x_n > \varepsilon\} \cap B_1,$$

for some $\varepsilon < \varepsilon_0$) then $\partial\Omega^+ \cap B_{1/2}$ si the graph of a Lipschitz function (and furthermore $D_\delta u \geq 0$ for any δ in a cone $\alpha(\delta, e_n) \leq \delta_0$).

b) If $D_\delta u$ is ≥ 0 for any δ such that $\alpha(\delta, e_n) \leq \delta_0$ then $\partial\Omega^+$ is a $C^{1,\alpha}$ surface.

Proof. According to the discussion we held at the beginning of these lectures, we will prove this result through a Harnack type inequality for free boundary problems.

The basic result should be the following: Let $u_1 \leq u_2$ be two solutions to the same free boundary problem, (in B_1). Assume further that at X_0, with $|X_0| < 1-\mu$, $d(X_0, \partial\Omega^+) > \mu$, we have

$$u_2(X_0) > u_1(X_0) + \varepsilon .$$

Then, on $B_{1/2}$

$$d(\partial\Omega^+(u_2, \partial\Omega^+(u_1)) \geq C\varepsilon .$$

Since this theorem in full generality is somewhat intricate, we will discuss here mainly part b), that is how a Lipschitz free boundary is really a $C^{1,\alpha}$ surface.

The main tool, both in the general as in the particular case, is a continuous perturbation lemma that allows us to do the following comparison argument:

We want to construct a continuous family of subsolutions, starting roughly at u_1, and lifting continuously the data near X_0. (By Harnack's inequality $u_2 - u_1 |B_{\mu/2}(X_0) \geq C\varepsilon$.

We further want, as we do so, the free boundary to advance continuously (in $B_{1/2}$) at a strictly positive speed, that is an amount proportional to the order of the perturbation. If we manage to do so, proving simultaneously that these free boundaries can never touch the free boundary of u_2 our Harnack inequality would be complete.

The perturbation argument is based in this observation.

Lemma: Let u be harmonic in B_1, and let ϕ be a (small) positive function satisfying $\phi\Delta\phi \geq C|\nabla\phi|^2$ (for C large enough depending on dimension).

Then

$$v(x) = \sup_{B_{\phi(x)}} u$$

is subharmonic.

Proof. Let $\phi(X) = \phi(0) + \nabla\phi \cdot X + \frac{1}{2}\phi_{ij}X_iX_j$.

We will prove that (at zero), $⨍_{\partial B_\varepsilon} (v(x) - v(0)) \geq - o(\varepsilon^2)$.

We assume that $v(0) = 0 = u(\phi(0)e_n)$.

In particular $\nabla u(\phi(0)e_n) = \lambda e_n$ with $\lambda \geq 0$. We may then estimate from below $v(X) \geq u(X + \phi(X)\nu(x))$.

We write $\nu(x) = e_n + \mu(x)$. Assume for simplicity

$$\nabla\phi = \phi_1 e_1 + \phi_n e_n = \alpha e_1 + \beta e_n$$

define γ by $(1+\gamma)^2 = (1+\beta)^2 + \alpha^2$ and chose

$$\nu(X) = \{e_n + \frac{1}{\phi(X)}[(\beta X_1 - \alpha X_n)e_1 + \sum_{1<j<n} \gamma X_j e_j]\}(1 + 0[\frac{|\nabla\phi|}{|\phi|})^2(|x|^2)).$$

so that $|\nu(X)| = 1$ (0 depends only on dimension). Then if we develop $x + \phi(x)\nu(x)$ up to $o(|x|^2)$ - order terms - we get

$$X + \phi(X)\nu(X) = X + \phi(X)e_n + [\beta X_1 - \alpha X_n]e_1 + \gamma(\Sigma X_j e_j) + 0(\frac{|\nabla\phi|^2}{\phi})\ |x|^2.$$

(with 0 depending only on dimension). We also write $\phi(X)e_n = \phi(0)e_n + (\alpha X_1 + \beta X_n)e_n + \frac{1}{2}\phi_{ij}X_iX_je_n$ and gather the terms of same order

$$X^* = X + \phi(x)\nu(x) = \phi(0)e_n + \{[(1+\beta)X_1 - \alpha X_n]e_1\} + (1+\gamma)\ \Sigma X_j e_j \cdot$$
$$+ [\alpha X_1 + (1+\beta)X_n]e_n\} + \{(\frac{1}{2}\sum_{ij}\phi_{ij}X_iX_j)e_n + 0(\frac{|\nabla\phi|^2}{\phi})|x|^2\}.$$

or

$$X^* = \phi(0)e_n + LX + Q(X) = \bar{X} + Q(X)$$

Then

$$\fint_{\partial B_\varepsilon} v(X) = \fint u(X_0)\,dx = \fint u(X^*) - u(\bar{X})\,dx + \fint u(\bar{X})\,dx$$

Bu
But

$$LX = \begin{vmatrix} 1+\beta & & & & \alpha \\ & 1+\gamma & & & \\ & & \ddots & & \\ & & & 1+\gamma & \\ -\alpha & & & & 1+\beta \end{vmatrix} X =$$

$$= 1+\gamma \begin{vmatrix} \cos\theta & & & & \sin\theta \\ & 1 & & & \\ & & \ddots & & \\ & & & 1 & \\ -\sin\theta & & & & \cos\theta \end{vmatrix} X$$

is a rotation plus a homotety.

Hence $\fint u(\bar{X})\,dx = 0.$ About

$$\fint_{\partial B_\varepsilon} u(X^*) - u(\bar{X})\,dx = \frac{1}{2} \fint_{\partial B_\varepsilon} |\nabla u((\phi(0)e_n)|\phi_{ij}(0)X_iX_j + |\nabla u|\frac{|\nabla\phi|^2}{\phi} 0(|X|^2) + 0(|X|^3)$$

(Recall that, at $\phi(0)e_n, \cdot|\nabla u| = \nabla u \cdot e_n$). That is

$$\fint_{\partial B_\varepsilon} v(x) \geq |\nabla u|[\Delta\phi - 0\,(\frac{|\nabla\phi|^2}{\phi})]\varepsilon^2 + 0(\varepsilon^3)$$

and the lemma is complete.

Let us now show how we construct with the aid of this lemma the family of supersolutions in case b).

That is, we know that u, <u>in</u> B_1, <u>satisfies</u> $D_\delta u \geq 0$ <u>for any</u> δ <u>in</u> $\{\alpha(\delta,e_n) \leq \tau\}$. We want to show that <u>in</u> $B_{1/2}$, <u>for some new vector</u> e_n', u <u>satisfies</u> $D_\delta u \geq 0$ <u>for any</u> δ in $\{\alpha(\delta,e_n') \leq \tau'\}$ with $\alpha(e_n',e_n) \leq (\tau' - \tau)$ and

$$(\pi/2 - \tau') \le \mu(\pi/2 - \tau)$$

with $\mu < 1$ independent of τ.

As in our discussion of the De Giorgi method for minimal surfaces, the new "vertical" direction should be chosen carefully.

In our case, we chose the direction by that of ∇u at say $1/2e_n$. That is assume that $\nabla u(\frac{1}{2} e_n) = ae_n + be_1$ $(a,b > 0)$. Since $D_\delta u \ge 0$ for any admissible δ,

$$\alpha(\nabla u(\frac{1}{2} e_n), e_n) \le \pi/2 - \tau_0 .$$

In other words, if π denotes the plane perpendicular to $\nabla u(\frac{1}{2} e_n)$ and $\Gamma(e_n,\tau)$ the cone

$$\{\delta/\alpha(\delta,e_n) \le \tau\},$$

π cannot intersect $\Gamma(e_n,\tau)$ and

$$\frac{D_\delta u(\frac{1}{2} e_n)}{D_n u(\frac{1}{2} e_n)} \sim d(\delta,\pi),$$

and, by Harnack inequality

$$\frac{D_\delta u(x)}{D_n u(x)} \sim d(\delta,\pi)$$

in a whole ball $B_\mu(\frac{1}{2} e_n)$.

With all these observations at hand, we will now develop a comparison principle among u and its translations: For that purpose we rewrite the fact

$$D_\delta u \ge 0, \quad \forall\ \delta \in \Gamma(e_n,\tau)$$

as

$$\sup_{B_{\varepsilon\cos\tau}} u(X - \varepsilon e_n) \le u(x)$$

for any ε, any x.

Choose now any direction δ in $\Gamma(e_n,\tau)$ and let us perform a further translation, $\varepsilon K\delta$.

It is still true that

$$\sup_{B_{\varepsilon\cos\tau}} u(X-\varepsilon(e_n+K\delta) \le u(X)$$

for any X in B_1. But from the previous discussion abou $\frac{D_\delta}{D_n}$ in $B_\mu(\frac{1}{2}e_n)$ and noting that from the non degeneracy, $D_n(\frac{1}{2}e_n) \ge \lambda_0 > 0$, on $B_\mu(1/2\ e_n)$, we have the further inequality

$$\sup_{B_{\varepsilon\cos\tau}} u(X-\varepsilon(e_n+K\delta)) \le u(X) - K\varepsilon d(\delta,\pi).$$

We now construct our family of subsolutions $v_t(x)$ starting with

$$v_0(X) = \sup_{B_{\varepsilon\cos\delta}} u(X-\varepsilon(e_n+K\delta)),$$

and continuing with

$$v_t(X) = \sup_{B_{\varepsilon\phi_t(X)}} u(X-\varepsilon(e_n+K\delta))$$

with $\phi_0 = \cos\delta$ and ϕ_t a proper choice of perturbation as to apply our auxiliary lemma on subharmonic functions.

That is, ϕ_t must satisfy

a) $\phi_t\Delta\phi_t \ge C|\nabla\phi_t|^2$ on $B_{1/2}(e_n)$

b) $\cos\tau \le \phi_t \le \cos\tau + t$

c) $\phi_t|_{B_1/2(0)\setminus[B_\mu(1/2\ e_n)\ \cup\ B_{3/8}(0)]} = \cos\tau$, and

$\phi_t|_{\partial B_{1/2}(0)\ \cap\ B_{\mu/2}(1/2)e_n)} = \cos\tau + t$

d) $\phi_t|_{B_{1/4}(0)} \sim \cos\delta + Ct \quad (0 < C < 1)$

e) $|\nabla\phi_t| \le Ct$.

It is not hard to construct such a ϕ, since the condition

$$\phi\Delta\phi \ge C\ |\nabla\phi|^2$$

simply means ϕ^{1-C} is superharmonic. We now follow the family of functions v_t. From our auxiliary lemma v_t is subharmonic whenever $v_t > 0$ or $v_t < 0$, therefore v_t remains smaller or equal to u unless $\partial\Omega^+(v_t)$ comes into contact with $\partial\Omega^+(u)$, say for $t = t_0$ at $X = X_0$.

From the definition of v as a sup on B_ϕ, it is not hard to see that $\partial\Omega^+(v)$ (and hence $\partial\Omega^+(u)$) have a tangent ball by the inside ad X_0. Therefore u satisfies at X_0 the asymptotic behavior

$$u = \alpha\langle X - X_0,\nu_0\rangle^+ - \beta\langle X - X_0,\nu_0\rangle^- + o(|X - X_0|),$$

with $\alpha = G(\beta,X_0,\nu_0)$.

On the other hand, also by the construction of v, $v(X_0) = u(X_1)$ at a (different) free boundary point where $\sup_{B_{\varepsilon\phi_t}(X_0)} u$ is attained.

At such a point $\partial\Omega^+$ has a tangent ball from outside and hence also (near X_1)

$$u(X) = \bar{\alpha}\langle X - X_1,\nu_1\rangle^+ - \bar{\beta}\langle X - X_1,\nu_1\rangle + o(|X - X_1|)$$

with $\bar{\alpha} = G(\bar{\beta},X_1,\nu_1)$. It follows that

$$v(X) \geq \bar{\bar{\alpha}}\langle X - X_0,\nu_0\rangle^+ - \bar{\bar{\beta}}\langle X - X_0,\nu_0\rangle^- + o(|X - X_0|)$$

with

$$\bar{\bar{\alpha}} \geq (1 - \varepsilon|\nabla\phi_t|)\bar{\alpha}$$

$$\bar{\bar{\beta}} \geq (1 - \varepsilon|\nabla\phi_t|)\bar{\beta}$$

$$|\nu_0 - \nu_1| \leq \varepsilon|\nabla\phi_t|$$

$$|X_0 - X_1| < |\phi_t|.$$

Finally, if t is a small multiple of $Kd(\delta,\pi)$, a comparison theorem for harmonic functions in Lipschitz domains (see [D]), asserts that, on $\Omega^+(u)$, $v_t \leq [1 - \varepsilon Kd(\delta,\pi)]u$ so we have

$$\bar{\alpha}(1 - \varepsilon|\nabla\phi_t|) \le \bar{\bar{\alpha}} \le \alpha(1 - \varepsilon K d(\delta,\pi)\bar{\beta}(1 - \varepsilon|\nabla\phi_t|) \le \bar{\bar{\beta}} \le \beta \ .$$

Given the regularity, and strict monotonicity hypothesis on G, this is a contradiction for t a small enough multiple of $d(\delta,\pi)$ (or what is the same, K large).

That is $v_t \le u$ on $B_{1/4}(0)$ for $t \le d(\delta,\pi)$. That is

$$\sup_{B_{\varepsilon(\cos\tau + d(\delta,\pi))}} u(X - \varepsilon(e_n + K\delta) \le u(X)$$

for any ε, for any X. A simple calculation shows that u is then monotone in a new cone, with the properties required above.

By repeating this argument inductively, we prove that on $B_{4^{-k}}$, $D_\delta u \ge 0$ for any δ in a cone $\Gamma_k = \Gamma(\tau_k, e_n^k)$ with $(\pi/2 - \tau_k) \le \mu^k$ and $\Gamma_{k+1} \supset \Gamma_k$. This readily implies that $\partial\Omega^+$ is $C^{1,\alpha}$.

To show that "flatness" implies Lipschitz (that is part a) of our theorem) the argument is similar but ε is not allowed, (initially) to go to zero, that is ε is allowed to diminish from step to step, (after normalization to B_1) but the aperture of the cone is also forced to diminish, ending up in a Lipschitz surface with Lipschitz constant smaller than that expected from its original flatness. A detailed proof will appear elsewhere.

As a last remark, let us point out that for ϕ_k = constant our perturbation maps level surfaces of u into parallel surfaces, that is, it corresponds to normal perturbations of minimal surfaces.

If one tries to apply this type of perturbations to prove initial regularity of minimal surfaces (flatness implies Lipschitz or Lipschitz implies $C^{1,\alpha}$) it appears that the original Harnack inequality is still necessary.

Bibliography

[A-C] H. W. Alt and L. A. Caffarelli, Existence and regularity for a minimum problem with free boundary, J. Reine Angew. Math. 105, 105-144 (1981).

[A-C-F] Alt-Caffarelli and Friedman, Variational problems with two phases and their free boundaries, T.A.M.S. Vol. 282, Nr.2, 1984, pp. 431-461.

[C-K] L.A. Caffarelli and D. Kinderlehrer, Potential methods in variational inequalities, J. Anal. Math. 27, 285-295 (1980).

[C-D-P] Colombini-De Giorgi-Piccinini, Frontiere orientate di misura minima e questioni collegate. Pisa (1972).

[D] B. Dahlberg, Estimates for harmonic measure. Arch. Rat. Mech. Anal. 65 (1977), 278-288.

[F] J. Frehse, On the regularity of the solution to a second order variational inequality, Boll. Unione Mat. Italiana 6 (4), 312-315 (1972).

[L-S] H. Lewy and G. Stampacchia, On the regularity of the solution of a variational inequality, Commun. Pure Appl. Math. 22, 153-188 (1969).

[S] D. Schaeffer, Some examples of singularities in a free boundary, Ann. Sc. Norm. Pisa 4 (4), 131-144 (1977).

MINIMAL FOLIATIONS ON A TORUS

Jürgen Moser

Abstract: These lectures are concerned with foliations of codimension one on a torus T^m whose leaves are minimals of a nonlinear variational problem. For $m = 2$ such foliations correspond to an invariant torus of a Hamiltonian system of two degrees of freedom as they occur in stability theory. The recently studied invariant Aubry-Mather-sets of monotone twist mappings have as analogue a "lamination" of the torus.

This study of minimal foliations, motivated in part by the theory of dynamical systems, depends strongly on tools of variational methods and nonlinear elliptic differential equations. It is the aim to discuss the connection between the mechanical problems and the higher-dimensional variational problems. In particular, we will describe a stability theorem for foliations generalizing the invariant curve theorem to partial differential equations.

Table of Contents:

I am indepted to V. Bangert for critical remarks as well as his advice and to M. Struwe and E. Zehnder for improvements of the manuscript.

I) Description of the Background and of the Problems

1. Minimal foliations

a) We will consider foliations of codimension one on a compact manifold M, and restrict ourselves to the case of a torus $M = T^m$, since it will be important for us that the fundamental group is commutative. A smooth foliation consists of a continuous one parameter family of smooth hypersurfaces covering M simply. These hypersurfaces are called leaves of the foliation. Locally these leaves can be represented as the level surfaces of a smooth function without critical points. For $m = 2$ these leaves are curves which can be viewed as solution curves of a vectorfield - or more precisely as integral curves of a line field, since these curves generally do not admit a consistent orientation. This is exemplified by the level curves of

(1.1) $$e^{x_1} \sin^2(\pi x_2)$$

on the two torus $T^2 = \mathbb{R}^2/\mathbb{Z}^2$.

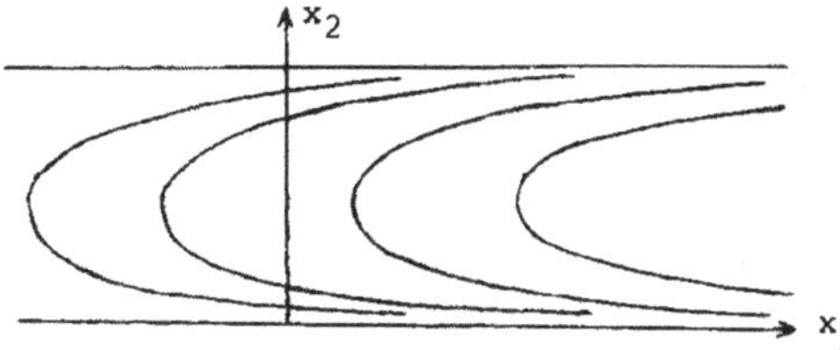

A foliation can be viewed as a generalization of a vector field the leaves corresponding to the orbits. For codimension one it can also be described by the field of normals. The leaves can be compact (as $x_2 = 0$ in the above example) or non-compact. We will primarily be interested in foliations with non-compact leaves. As example we consider the family of parallel hyperplanes

(1.2) $$\sum_{\nu=1}^{m} \alpha_\nu x_\nu = \text{const} \; ; \quad \bar{\alpha} = (\alpha_1, \ldots, \alpha_m) \neq 0 \, .$$

They give rise to a foliation F^0_α of T^m. It consists of compact leaves (subtori) if and only if $\bar{\alpha}$ is the multiple of a rational vector. On the other hand, every leaf is dense on T^m if and only if the line $N = \{\lambda\bar{\alpha}; \lambda \in \mathbb{R}\}$ normal to the foliation meets no lattice point except (0), i.e.

(1.3) $$N \cap \mathbb{Z}^m = (0) \, .$$

For $m = 2$ these statements correspond to the fact that on a torus $T^2 = \mathbb{R}^2/\mathbb{Z}^2$ a line $\alpha_1 x_1 + \alpha_2 x_2 = 0$ corresponds to a dense curve on T^2 if and only if its slope $-\alpha_2/\alpha_1$ is irrational and to a circle if and only if its slope is rational (or infinity).

b) We call a foliation <u>minimal</u> with respect to a variational problem if every leaf minimizes this variational problem when viewed on the covering space $\mathbb{R}^m$. For example, if the functional is given by the $(m-1)$-dimensional area of the leaf with respect to a metric

(1.4) $$ds^2 = \sum_{\nu,\mu=1}^{m} g_{\nu\mu}(\bar{x})\, dx_\nu dx_\mu$$

with $g_{\nu\mu}(\bar{x}+\bar{j}) = g_{\nu\mu}(\bar{x})$ for all $\bar{j} \in \mathbb{Z}^m$, then the leaves have to be minimal surfaces in $\mathbb{R}^m$. For the flat metric

$$ds_o^2 = \sum_{\nu=1}^{m} dx_\nu^2$$

all affine hyperplanes are clearly minimal surfaces and therefore F_α^o given by (1.2) is an example of a minimal foliation with respect to the area-functional of ds_o^2.

Are there such minimal foliations with respect to another metric ds^2? We give here a result illustrating a typical phenomenon to be discussed. For this purpose we call a foliation F^o, which is minimal w.r. to the area-functional of ds_o^2, <u>stable</u> if for any nearby metric ds^2 (nearby in C^∞-topology) there exists a foliation F, which is minimal with respect to ds^2 and diffeomorph to F^o. This means that there exists a smooth diffeomorphism ψ (near the identity) of T^m taking the leaves of F^o into those of F.

Are there stable foliations? One convinces oneselves easily, that a family of minimal subtori, say $x_m = \text{const}$ will disintegrate into a family of only finitely many such tori under perturbation of the metric, showing that the foliation (1.2) with $\bar{\alpha} = (0,0,\dots,0,1)$ will be unstable. On the other hand, if the leaves of (1.2) are dense there is a better chance for stability. We strengthen the condition (1.3) and require that the normal N does not come "too close" to the lattice $\mathbb{Z}^m$, by imposing the Diophantine condition

(1.5) $$\sum_{1\le\nu<\mu\le m} (\alpha_\nu j_\mu - \alpha_\mu j_\nu)^2 \ge |\bar{\alpha}|^2 c_o^{-1} |\bar{j}|^{-2\tau} \quad \text{for all } \bar{j} \in \mathbb{Z}^m \setminus (0)$$

for some positive constants τ, c_o. Incidentally, for almost all vectors $\bar{\alpha} \in \mathbb{R}^m$

such constants c_o, τ do exist. By the way, this condition does <u>not</u> imply that the α_ν are rationally independent; for $m \geq 3$ one could even have, for example, vectors $\bar{\alpha}$ with $\alpha_2 = 1$, $\alpha_3 = \ldots = \alpha_m = 0$, satisfying (1.5).

<u>Theorem 1.1.</u> If $\bar{\alpha}$ satisfies the condition (1.5) then the foliation F^o_α. given by (1.2), and minimal with respect to ds^2_o is stable.

This result shows that the question of stability is related to number theoretical conditions on the normal vector $\bar{\alpha}$. On the other hand, if $\bar{\alpha}$ is a rational vector then F^o_α is definitely not stable.

For $m = 2$ these minimal foliations correspond to fields of geodesics on a torus T^2. For a flat torus T^2 all the geodesics are straight lines on $\mathbb{R}^2$, and for each direction $\bar{\alpha} = (\alpha_1, \alpha_2) \in \mathbb{R}^2 \setminus (0)$ one has a minimal foliation normal to $\bar{\alpha}$. It is a wellknown that any geodesics which is embedded in a field of extremals, such as the leaves of a foliations, are globally minimal, in particular, have no conjugate points.

According to a theorem by E. Hopf a metric on the torus, for which all geodesics are free from conjugate point, is necessarily flat. This shows that for arbitrary metrics not all geodesics are members of a minimal foliation.

c) In the following we will restrict ourselves to foliations whose leaves are representable as the graphs of a function. For this purpose we set

$$m = n+1\ , \quad \bar{x} = (x_1, x_2, \ldots, x_{n+1}) \in \mathbb{R}^{n+1}$$

(1.6)

$$x = (x_1, x_2, \ldots, x_n) \in \mathbb{R}^n$$

and ask that the leaves on the covering space $\mathbb{R}^m$ are representable as graphs

$$x_{n+1} = u(x)\ .$$

The above example (1.1) shows that not every foliation admits such a representation. This rather ungeometrical restriction is related to the non-parametric character of the variational problem:

(1.7) $$\int F(x,u,u_x)\,dx \quad ; \quad dx = dx_1 \wedge \ldots \wedge dx_n\ .$$

Here we assume that

(1.8)

(i) $F = F(\bar{x},p) \in C^{2,\varepsilon}(T^{n+1} \times \mathbb{R}^n)$

i.e. F has period 1 in $x_1,\dots,x_{n+1}$.

(ii) There exists a $\delta \in (0,1]$ such that

$$\delta|\xi|^2 \le \sum_{\nu,\mu=1}^{n} F_{p_\nu p_\mu} \xi_\nu \xi_\mu \le \delta^{-1}|\xi|^2$$

for all $\xi \in \mathbb{R}^n$ (Legendre condition).

(iii) Quadratic growth: There exist positive constants δ , c_1,c_2 such that

$$\delta|p|^2 \le F(\bar{x},p) \le \delta^{-1}|p|^2 + c_1$$

$$\sum_{j=1}^{n+1} |F_{x_j}(\bar{x},p)| \le c_2(|p|^2+1)$$

for all $\bar{x} \in \mathbb{R}^{n+1}$, $p \in \mathbb{R}^n$.

The periodicity condition i) means that the variational problem is defined on the torus T^{n+1} . The conditions (1.8), which are weaker than those of [17], suffice for estimates of minimals (see [7]).

The typical example is a quadratic polynomial

$$F = (a(\bar{x})p,p) + 2(b(\bar{x}),p) + c(\bar{x})$$

with coefficients of period 1 in $x_1,x_2,\dots,x_{n+1}$ and a positive symmetric matrix $a(\bar{x})$. A special case is

(1.9) $$F = \frac{1}{2}|p|^2 + V(\bar{x}) ,$$

the two terms corresponding to kinetic and potential energy. We call $u = u(x)$ an <u>extremal</u> of the variational problem if it is solution of the Euler equation

(1.10) $$\sum_{\nu=1}^{n} \partial_{x_\nu} F_{p_\nu}(x,u,u_x) = F_{x_{n+1}}(x,u,u_x) .$$

Instead of considering the foliation on T^{n+1} we look at its lift, a foliation on $\mathbb{R}^{n+1}$, and have to take into account that this foliation is invariant under the action of $\mathbb{Z}^{n+1}$.

<u>Definition 1.2.</u> A $\mathbb{Z}^{n+1}$-invariant F-minimal foliation is defined by a continuous function $u : \mathbb{R}^n \times \mathbb{R} \to \mathbb{R}$ with the following properties:

i) For each $\lambda \in \mathbb{R}$ the function $u = u(x,\lambda)$ is a solution of the Euler equation for all $x \in \mathbb{R}^n$.

ii) For each $x \in \mathbb{R}^n$ the mapping $\lambda \to u(x,\lambda)$ is a homeomorphism of $\mathbb{R}$ onto $\mathbb{R}$,, in particular, $u(x,\lambda) < u(x,\lambda')$ for $\lambda < \lambda'$.

iii) The foliation given by the leaves $\{x_{n+1} = u(x,\lambda) , \lambda \in \mathbb{R}\}$ is invariant under the $\mathbb{Z}^{n+1}$-action.

It is a consequence of regularity theory that the leaves are in C^2, i.e. $u(\cdot,\lambda) \in C^2$, even if the u are assumed to be only weak $H^{1,2}_{loc}$-solutions of (1.10). However, we require only continuous dependence on λ.

Because of the smoothness of the leaves of a minimal foliation the unit normal (with position x_{n+1}-component) is well-defined and depends continuously on its base point. It is remarkable that this normal vector is always Lipschitz-continuous. On the other hand, Lipschitz continuity actually is optimal as this example shows: If $n = 1$ and

$$F = \frac{1}{2}(u_x^2 - \frac{1}{2\pi}\cos 2\pi u)$$

the Euler equation becomes

$$u_{xx} = \frac{1}{2}\sin 2\pi u \ .$$

For this problem a minimal foliation is given by the equation

$$u_x = \frac{1}{\pi}|\sin \pi u|$$

which has merely a Lipschitz-continuous normal, while the leaves given by

$$\tan\frac{\pi}{2}u = ce^x \quad \text{for} \quad c \geq 0 \ ; \ 0 \leq u < 1$$

are obviously analytic.

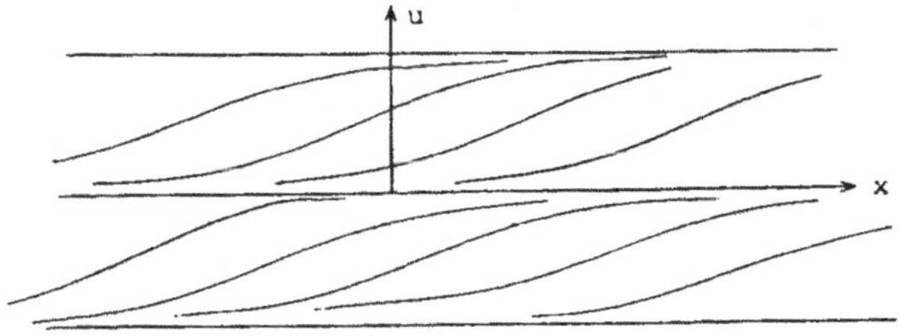

It would be more appropriate to speak of an extremal foliation, since we require the leaves merely to be extremal of the variational problem (1.7). However, it follows that any leaf of such a foliation is automatically minimal in the following global sense:

Definition 1.3. A function $u \in H^{1,2}_{loc}(\mathbb{R}^n)$ is called a minimal of (1.7) if

(1.11) $$\int_{\mathbb{R}^n} (F(x,u+\phi,u_x+\phi_x) - F(x,u,u_x))\,dx \geq 0$$

for all $\phi \in H^{1,2}_{comp}(\mathbb{R}^n)$.

Clearly, not every extremal is minimal while, of course, every minimal is an extremal. Indeed, for $n = 1$ it is clear that any extremal with conjugate points is not minimal. In general, the set of minimals forms a proper subset of the set of extremals. For foliations, however, this difference between extremals and minimals disappears.

We will not give a complete proof of the above statement that an "extremal" foliation is automatically a minimal foliation, which justifies our terminology, but indicate the ideas: If $u = u(x,\lambda_0)$ is a leaf of a foliation which is extremal but not minimal, then there exists a ball $B \subset \mathbb{R}^n$ and a $v \in H^{1,2}(\mathbb{R}^n)$ such that $v = u$ outside B and

$$\int_B F(x,v,v_x)dx < \int_B F(x,u,u_x)dx .$$

We may take v to be the minimal of the integral on the right with the boundary condition $v = u$ on ∂B, and so v satisfies the Euler equation in B. Since $v \not\equiv u$ in B we have either

$$\max_{x\in B}(v(x)-u(x,\lambda_0)) > 0 \quad \text{or} \quad \min_{x\in B}(v(x)-u(x,\lambda_0)) < 0 .$$

We assume we are in the first case and consider the function

$$\mu(\lambda) = \max_{x\in B}(v(x)-u(x,\lambda)) .$$

which for $\lambda \to +\infty$ tends to $-\infty$ hence assumes the value 0 for some $\lambda = \lambda^* > \lambda_0$. Now this implies that $v(x) \leq u(x,\lambda^*)$ for all $x \in B$ with equality at some interior point of B. In other words $u(x,\lambda^*) - v(x)$ assume its minimum, namely 0, at some interior point of B. Since this difference satisfies an elliptic partial differential equations this contradicts the maximum principle.

Thus the leaves of a minimal foliation always are minimals in the sense of Definition 1.3. The leaves of a foliation have another property which will be important for us, namely that they have no selfintersection on the torus. For the lifted foliation this means the following: If $x_{n+1} = u(x)$ is a leaf of the $\mathbb{Z}^{n+1}$-invariant foliation, so is the translated leaf

$$x_{n+1} + j_{n+1} = u(x+j) , \qquad j \in \mathbb{Z}^n .$$

We say u has <u>no selfintersection</u> (on T^{n+1}) if for every fixed $(j,j_{n+1}) \in \mathbb{Z}^{n+1}$

$$u(x+j) - j_{n+1} - u(x)$$

does not change the sign, i.e. if for all $x \in \mathbb{R}^n$ we have

(1.12) $$u(x+j) - j_{n+1} - u(x) > 0 \quad \text{or} \quad \equiv 0 \quad \text{or} \quad < 0 .$$

The case that this difference is ≥ 0 and has some zero implies by the maximum principle that it vanishes identically. Thus we can equivalently say, u <u>has no selfintersection (on T^{n+1}) if and only if the set of translates</u>

$$\tau_{\bar{j}}u := u(x+j) - j_{n+1} \quad ; \quad \bar{j} \in \mathbb{Z}^{n+1}$$

<u>is totally ordered.</u>

We record for later: The leaves of a minimal foliation are a) minimal in the sense of (1.11) and b) have no selfintersections in the sense of (1.12).

2. Problems, Phenomena, Motivations

a) In the next lecture we will see that we can associate with any minimal foliation an asymptotic normal vector $\bar{\alpha}$, as in the first example. Since we restrict ourselves to foliations whose leaves are graphs the last component of this vector is $\neq 0$ and can be normalized to be -1, i.e.

$$\bar{\alpha} = (\alpha_1 \ldots \alpha_n, -1) \quad ; \quad \alpha = (\alpha_1, \ldots, \alpha_n) \; .$$

The vector $\alpha \in \mathbb{R}^n$ will be called the slope vector. A trivial example of a minimal foliation for

$$\int u_x^2 \, dx$$

is given by

$$x_{n+1} - (\alpha, x) = \text{const} \; . \tag{2.1}$$

We will see that for any minimal foliation whose leaves are graphs $x_{n+1} = u(x)$ there exists a unique slope vector $\alpha \in \mathbb{R}^n$ so that

$$\sup_x |u(x) - (\alpha, x)| < \infty \; . \tag{2.2}$$

This vector is independent of the individual leaf and is defined by the above condition.

b) Thus every minimal foliation considered has such a slope vector $\alpha \in \mathbb{R}^n$. We are led to several questions: 1) Given a variational problem (1.7) can one, for any slope vector α, find a corresponding minimal foliation? 2) Are such foliations stable under perturbation of the integrand F of the variational problem?

It turns out that already the first question has a negative answer! However, if the concept is appropriately generalized the answer is positive: For every $\alpha \in \mathbb{R}^n$ there exists a minimal "lamination". It can be viewed as a foliation of a certain subset of the torus, and this subset is a Cantor set (unless it is the whole torus), uniquely associated to F and α. This leads to the question: Under which circumstances do we have foliations covering the whole torus and when does one have a lamination on a Cantor set?

We will describe the minimal foliation by a certain function which has to satisfy a nonlinear partial differential equation (see lecture IV). A smooth solution of these equations will correspond to a smooth foliation, and a "lamination" (which we have not defined so far) to a weak solution possessing discontinuities.

Finally, we ask how such generalized foliations depend on parameters. For example, take

$$(2.3) \qquad F(x,u,u_x) = \frac{1}{2}|u_x|^2 + \lambda V(x,u)(1+|u_x|^2)^{1/2}$$

with $V(x,x_{n+1})$ of period 1 in all variables and smooth. If $\bar{\alpha} = (\alpha,-1)$ satisfies the condition (1.5) then one has indeed smooth foliations for small $|\lambda|$ but for large $|\lambda|$ they may disintegrate or "break down", as examples show. Thus the solutions may loose smoothness as λ changes. On the other hand, no matter how large $|\lambda|$ there exist smooth minimal foliations, for certain α with $|\alpha|$ large enough. Thus this question is extremely subtle and some remarks in this direction will be discussed in lecture III.

c) This theory was motivated by problems in mechanics, in particular, the stability problem for dynamical systems. For systems of two degrees of freedom such a stability theory is based on the construction of two-dimensional invariant tori (KAM theory). The phenomenon of break-down of stability is related to the disintegration of invariant tori. The understanding of these phenomena has been advanced considerably by the important work [13] of J. Mather who studied discrete systems, the iterates of area-preserving mappings, called monotone twist mappings. At the same time Aubry et al studied simple models for the motion of electrons in a one-dimensional crystal. The minimum energy configurations of Aubry correspond to the orbits on invariant sets for monotone twist mappings found by Mather (for a review, see [1]).

This theory of Aubry and Mather is limited strictly to 2-dimensional mappings or one-dimensional crystals, both corresponding to Hamiltonian systems of two degrees of freedom. Efforts to generalize their theory to more than two degrees of freedom have failed. However, replacing the one-dimensional orbits of Hamiltonian systems by surfaces of codimension one leads to a different higher dimensional generalization which we discuss here. The leaves of a minimal foliation can be viewed as the minimal configurations in Aubry's theory when considered in an n-dimensional lattice.

The connection of the above minimal foliations to the theory of Aubry and Mather has been discussed in [16]. It corresponds to the special case $n=1$, i.e. foliation of a two-dimensional torus. To be precise, their theory refers to discrete systems while we will describe the continuous analogue, which was studied by J. Denzler [6]. Therefore we turn now to the case $n=1$ and discuss the relevance of minimal foliation for a simple stability problem.

3. Connection with a stability problem

a) A nonlinear pendulum is described by the differential equation

$$(3.1) \qquad \frac{d^2x}{dt^2} = g(t)\sin(2\pi x)$$

where x is the angle describing the position of a mass-point and $g(t)$ the vertical force, which we allow to depend periodically on t with period 1. For constant $g(t)$ the solutions are well-understood because of the conservation of the energy: In that case all solution with the exception of the asymptotic orbits (separatrices) are periodic if the motion is viewed on the torus T^2 described by (t,x) modulo 1. In any case, for all solutions $x = x(t)$ the velocity $\dot{x}(t)$ is bounded for all $t \in \mathbb{R}$. We ask, whether this is also true for the case that the force $g(t)$ is a periodic function of t.

For our purposes we call a system

$$\frac{d^2x}{dt^2} = V_x(t,x) \tag{3.2}$$

with $V = V(t,x)$ smooth and of period 1 in t and x (e.g. $V = -\frac{1}{2\pi}g(t)\cos(2\pi x)$ for (3.1)) stable if for every of its solution

$$\sup_{t\in\mathbb{R}} |\dot{x}(t)| < \infty .$$

Is every system (3.2) stable, or can one pump the system up by a periodic force so that the angular velocity becomes unbounded? It turns out that every such system is stable in this sense. How to prove this?

The system (3.2) describes the Euler equation of the variational problem

$$\int F(t,x,\dot{x})\,dt \quad \text{with} \quad F(t,x,p) = \frac{1}{2}p^2 + V(t,x) . \tag{3.3}$$

It can also be written as a Hamiltonian system

$$\dot{x} = H_y \; ; \; \dot{y} = -H_x \quad \text{with} \quad H(t,x,y) = \frac{1}{2}y^2 - V(t,x) \tag{3.4}$$

which can be viewed as a vectorfield in the 3-dimensional phase-space $T^2 \times \mathbb{R}$. Our question is whether $y = \dot{x}$ is bounded for all t.

b) A sufficient condition for the boundedness of $y = y(t)$ for a solution is that it can be enclosed between two invariant tori given by two functions $w_\nu = w_\nu(t,x)$, $\nu = 1,2$ of period 1 in t,x by

$$y = w(t,x) , \quad w = w_1, w_2 .$$

We call such a torus invariant if the vectorfield (3.4) is tangential to it, so that the torus can be viewed as a set of orbits. If we write w as the x-derivatives of a function $S = S(t,x)$, i.e.

$$y = S_x(t,x) \tag{3.5}$$

then it defines an invariant torus if and only if S satisfies the Hamilton-Jacobi equation

$$S_t + H(t,x,S_x) = f(t)$$

with an arbitrary function $f = f(t)$. Adding a function of t to S (which does not change w) finding an invariant torus amounts to finding a solution of the Hamilton-Jacobi equation

(3.6) $$S_t + H(t,x,S_x) = 0$$

for which $S_x(t,x)$ has period 1 in t,x. It is well-known how to solve the Hamilton-Jacobi equation locally, or even the initial value problem, but this is a boundary value problem whose solution is very subtle.

Assume we have such a solution of (3.6) hence an invariant torus then the projection of the phase-space $T^2 \times \mathbb{R} \to T^2$ on the torus which maps (t,x,y) into (t,x) will take the orbits on the invariant torus into a family of curves on the (t,x)-torus. They are clearly solutions of the system

$$\dot{x} = S_x(t,x) \ , \ \dot{t} = 1$$

which defines a foliation of the torus. Since they also are solutions of the Euler equation (3.2) they represent a minimal foliation. Conversely, any minimal foliation, given by a smooth vectorfield

(3.7) $$\dot{x} = w(t,x) \ , \ \dot{t} = 1$$

on the torus defines an invariant torus $y = w(t,x)$ in the phase-space.

Thus invariant tori correspond to minimal foliations in the sense discussed above. The traditional notation has been adopted: $t,x,\dot{x}$ have to be replaced by x,u,u_x. It is a well-known result in the theory of ordinary differential equation (see [21]) that one can associate a rotation number α to such a system (3.7) on T^2. If α is irrational the solutions have the form

$$x = \theta + p(t,\theta) \ ; \ \theta = \alpha t + \text{const}$$

where p has period 1 in t,x. They are quasi-periodic with frequencies $1, \alpha$, and the frequency α corresponds to the slope vector of the foliation since

$$x(t) - \alpha t$$

is bounded.

Thus to solve our stability problem it is enough to show that every neighborhood of $y = \infty$:

$$N_\varepsilon = \{(t,x,y) \in T^2 \times \mathbb{R} \ ; \ |y| > \varepsilon^{-1}\}$$

contains an invariant torus. This can be translated into finding minimal foliation for arbitrary large frequencies $|\alpha|$.

c) We will come back to this problem in lecture III. Here we wanted to explain the correspondence of the concepts in the theory of ordinary and partial differential equations, respectively, illustrated by the table:

ODE (n = 1)	PDE (n > 1)
orbits	solutions of the Euler equations
invariant torus	minimal foliation
frequency, rotation #	slope vector or rotation vector
Aubry-Mather sets	minimal laminations

We point out, however,that there are essential differences: The concept of (dynamical) stability has no analogue for $n > 1$ since it is based on the initial value problem which is meaningless for elliptic partial differential equations. Also the canonical transformation theory basic for Hamiltonian system has no analogue for elliptic partial differential equations.

d) In the following lectures we will describe two ways to construct minimal foliations or laminations for a given slope vector α . In the next section we obtain the generalized foliation by fitting together individual leaves. In the last section we describe an alternate approach which yields the foliation as the minimal of a degenerate variational problem, which will be solved by regularization. Some of the results have been published in [17], but we will also describe the important result by Bangert [3] which asserts that for a given variational problem and given $\alpha \notin \mathbb{Q}^n$ the associated lamination is uniquely determined.

II) Construction of a generalized minimal foliation

4. Minimals without selfintersections

a) In the following two sections it is our goal to construct a minimal foliation by piecing together the leaves which make it up. The procedure is illustrated by the simple example of the foliation of parallel hyperplanes $x_{n+1} - (\alpha,x) = \text{const}$ which constitute a minimal foliation for the Dirichlet integral. If we take one of these leaves, say

$$x_{n+1} = u_o(x) = (\alpha,x)$$

we can obtain others by translation under the fundamental group $G = \mathbb{Z}^{n+1}$:

$$x_{n+1} = u_o(x+j) - j_{n+1} = u_o(x) + (\bar{\alpha},\bar{j})$$

where $\bar{\alpha} = (\alpha_1,\ldots,\alpha_n,-1)$ and $\bar{j} \in G$. If $\alpha = (\alpha_1,\ldots,\alpha_n)$ is not rational than the numbers $(\bar{\alpha},\bar{j})$, $\bar{j} \in G$ are dense in $\mathbb{R}$ and thus all leaves of the foliation can be approximated by the translates of one of them in this case. This exemplifies the procedure which we now follow, except that, in general, the translates of a leaf need not be dense on the torus, even if α is not a rational vector.

First we will show that one can associate to any $\mathbb{Z}^{n+1}$-invariant minimal foliation (Definition 1.2) a unique slope vector $\alpha \in \mathbb{R}^n$ so that any of its leaves $x_{n+1} = u(x)$ has a bounded distance from the hyperplane $x_{n+1} = (\alpha, x)$, i.e. that (2.2) holds. The question to be addressed is: Given a variational problem (1.7) and a vector $\alpha \in \mathbb{R}^n$ does there exist a F-minimal foliation corresponding to the slope vector α ?

In Section 1 we saw that the leaves of such a foliation have two basic properties: They are minimals in the sense of Definition 1.3 and they have no selfintersection on the torus. Therefore we will study now such minimals without selfintersections, derive a-priori estimates and compactness properties which will allow us to piece them together (Section 4) to a minimal foliation or lamination.

b) Functions without selfintersections: In the space $C(\mathbb{R}^n)$ of continuous functions $u = u(x)$ we consider the group action of $G = \mathbb{Z}^{n+1}$ by the translations

(4.1) $$(\tau_{\bar{j}} u)(x) = u(x+j) - j_{n+1} \quad \text{for } \bar{j} = (j, j_{n+1}) \in \mathbb{Z}^{n+1} .$$

Motivated by the discussion at the end of Section 1 we give the following

Definition 4.1: $u \in C(\mathbb{R}^n)$ has no selfintersections if the orbit $\{\tau_{\bar{j}} u, \bar{j} \in \mathbb{Z}^{n+1}\}$ is totally ordered, i.e. if for every $\bar{j} \in \mathbb{Z}^{n+1}$ one has for all $x \in \mathbb{R}^n$ either $(\tau_{\bar{j}} u)(x) - u(x) > 0$ or < 0 or $\equiv 0$.

Theorem 4.2: If $u \in C(\mathbb{R}^n)$ has no selfintersections then there exists a vector $\alpha \in \mathbb{R}^n$ so that

$$\sup_{x \in \mathbb{R}^n} |u(x) - (\alpha, x)| < \infty .$$

Moreover, with this α one has

(4.2) $$|u(x+j) - u(x) - (\alpha, j)| \leq 1$$

(4.3) $$(\alpha, j) - j_{n+1} > 0 \quad \text{implies} \quad u(x+j) - j_{n+1} - u(x) > 0 .$$

This theorem is a consequence of standard results on circle mappings (see [17]). We indicate the argument.

Consider the denumerable set

$$S_x = \{(\tau_{\bar{j}} u)(x) \ , \ \bar{j} \in \mathbb{Z}^{n+1}\}$$

and define for $k \in \mathbb{Z}^n$ the mapping $f^k: S_x \to S_x$ by translating x by k :

$$f^k: u(x+j) - j_{n+1} \to u(x+j+k) - j_{n+1} .$$

Then, since u has no selfintersections, one has

$$f^k(s) < f^k(s') \quad \text{for } s < s' \ , \ s, s' \in S_x$$

$$f^k(s+1) = f^k(s) + 1 .$$

Moreover, the mappings f^k , f^ℓ commute for $k, \ell \in \mathbb{Z}^n$.

For such a mapping $f = f^k$ one has the following facts, taken from Denjoy's theory on circle homeomorphisms (see [17], Appendix to Section 2):

α) The limit
$$\lim_{q\to\infty} \frac{f^{kq}(s)}{q} = \alpha(k) = \sum_{\nu=1}^{n} \alpha_\nu k_\nu$$

exists and is independent of s. This number $\alpha(k)$ is called the rotation number of f^k.

β) For the periodic function $f^k(s) - s$ one has
$$|f^k(s) - s - \alpha(k)| \le 1 \quad \text{for all} \quad s \in S_x .$$

γ) For any two integers p,q one has: $q\alpha(k) - p > 0$ implies $f^{kq}(s) - p - s > 0$ for all $s \in S_x$. In other words the points $\{f^{kq}(s);\, q \in \mathbb{Z}\}$ have the same ordering on the circle $\mathbb{R}/\mathbb{Z}$ as the points $\alpha(k)\cdot q$.

The statements β) and γ) translate into (4.2), (4.3). To prove the first statement of the theorem 4.2 we set

$$Q = \{x \in \mathbb{R}^n ,\, |x_\nu| \le \tfrac{1}{2}\} \tag{4.4}$$

and pick, for given $x \in \mathbb{R}^n$, a vector $j \in \mathbb{Z}^n$ so that $x+j = y \in Q$. Then the function $w(x) = u(x) - (\alpha,x)$ satisfies by (4.2)

$$|w(y) - w(x)| \le 1 .$$

Hence
$$|w(x) - w(0)| \le |w(y) - w(0)| + 1 \le \operatorname{osc}_Q w + 1 .$$

This implies the boundedness of $w(x)$ and, more precisely, the estimate

$$|u(x) - u(0) - (\alpha,x)| \le 1 + \operatorname{osc}_Q (u(x) - (\alpha,x)) . \tag{4.5}$$

Because of the connection with the rotation number of a circle mapping we will call the slope vector α also the <u>rotation vector</u>.

c) Now we assume that u is a minimal of the variational problem 1.7.

<u>Definition 4.3</u>: A function $u \in H^{1,2}_{loc}(\mathbb{R}^n)$ is called a minimal of (1.7) if (1.11) holds.

Making use of the work of Giaquinta and Giusti on Q-minimals it follows that these minimals are Hölder continuous functions. For this purpose one has to require only the condition (1.8) i), iii) but not ii). Adding this assumption (1.8) ii) one can establish that the minimals are in $C^2(\mathbb{R}^n)$ and satisfy the Euler equation (1.10). Therefore we have the classical maximum principle for such minimals available:

<u>Proposition 4.4</u>: If u,v are two minimals satisfying

$$u \le v$$

in an open domain $\Omega \subset \mathbb{R}^n$ then one has either $u < v$ or $u \equiv v$ in Ω .

The family of minimals is invariant under the translation $\tau_{\bar{j}}$ (see (4.1)). We now consider the class of minimals without selfintersections which we denote by $\mathfrak{M}$. The requirement of "no selfintersection" is rather restrictive, as the example of the Dirichlet integral $F = \frac{1}{2}u_x^2$ shows: In this case every harmonic function is minimal but only the linear functions belong to $\mathfrak{M}$. By Theorem 3.2 we can associate to $u \in \mathfrak{M}$ a vector $\alpha \in \mathbb{R}^n$ and the minimals corresponding to the α will be combined into the set $\mathfrak{M}_\alpha$, so

(4.6) $$\mathfrak{M} = \bigcup_\alpha \mathfrak{M}_\alpha$$

is a disjoint union.

Before deciding the existence question whether or not $\mathfrak{M}_\alpha \neq \phi$ we collect some a-priori estimate and compactness properties. For the proofs we refer to [17].

d) Theorem 4.5: There exists a constant c_1 depending on F such that for any $u \in \mathfrak{M}_\alpha$ the inequality

(4.7) $$|u(x+y) - u(x) - (\alpha,y)| \leq c_1 \sqrt{1+|\alpha|^2}$$

holds. Here c_1 is independent of α and u .

Moreover, for all $|\alpha| \leq A$ there exists positive constant γ,ε, depending on F and A , such that for all $u \in \mathfrak{M}_\alpha$ with $|\alpha| \leq A$ one has

(4.8) $$|u_x|_{C^\varepsilon} \leq \gamma .$$

The first inequality which is basic for the following expresses that the graph of u has a distance $\leq 2c_1$ from the hyperplane $x_{n+1} = (\alpha,x) + u(0)$. It can be viewed as an analogue of a theorem of Hedlund [10] for minimal geodesics on a two-dimensional torus. His work was based on Morse [15] on such geodesics on surfaces of higher genus.

The proof requires, according to (4.5), an estimate of $\operatorname{osc}_Q u$. It can be derived with the help of the work of Giaquinta and Giusti [7,8]. They showed that minimals (even Q-minimals) belong to the so-called de Giorgi class, for which pointwise estimates can be established (see Ladyshenshaya and Uraltseva [12]). For the proof of Theorem 4.5 see [17].

The most important consequence of Theorem 4.5 is a compactness property of the minimal: If $u_m \in \mathfrak{M}_{\alpha_m}, |\alpha_m| \leq A$ is a sequence of minimals, for which we can assume that $0 \leq u_m(0) < 1$ by subtracting an integer, then a subsequence converges in the C^1-topology to a function u . By C^r-topology we mean uniform convergence of the functions and its derivatives up to order r on all compact subsets of $\mathbb{R}^n$.

One verifies easily that the limit function u is a minimal again. Moreover, u has no selfintersections. Indeed, it is clear that for every $\bar{j} \in \mathbb{Z}^{n+1}$ the function $\tau_{\bar{j}}u-u$ does not change the sign, and on account of Proposition 4.4 applied to $v = \tau_{\bar{j}}u$ one has strict inequality unless $\tau_{\bar{j}}u \equiv u$.

This argument leads to the following theorem, where $\mathfrak{M}_\alpha/\mathbb{Z}$ is obtained from $\mathfrak{M}_\alpha$ by identifying u with $u + \text{integers}$, and where $\mathfrak{M}_A = \bigcup_{|\alpha| \le A} \mathfrak{M}_\alpha$:

Theorem 4.6: The set $\mathfrak{M}_A/\mathbb{Z}$ is compact with respect to the C^0-topology. Moreover. $\alpha = \alpha(u)$ is a continuous function on $\mathfrak{M}_A/\mathbb{Z}$ in this topology.

We note that on account of Theorem 4.5 a sequence $u_m \in \mathfrak{M}$ converge C^0-topology if and only if it converges in the C^1-topology.

e) We turn to the existence of an element in $\mathfrak{M}_\alpha$. This is an unusual problem since the underlying domain, namely $\mathbb{R}^n$, of u is non-compact and the vector α represents, in some sense, a boundary condition at infinity. We circumvent these difficulties if we consider rational vectors $\alpha \in \mathbb{Q}^n$. Then the $j \in \mathbb{Z}^n$ satisfying $(j,\alpha) \in \mathbb{Z}$ form a sublattice of finite index of $\mathbb{Z}^n$ and the functions in

(4.9) $$\mathfrak{M}_\alpha^{per} = \{u \in \mathfrak{M}_\alpha \mid \tau_{\bar{j}}u = u \quad \text{whenever} \quad (\bar{j},\bar{\alpha}) = 0\}$$

define a graph which is a subtorus of T^{n+1}. In fact, $u(x)-(\alpha,x)$ is a periodic function whose periods consist of the sublattice $\{j \in \mathbb{Z}^n \mid (j,\alpha) \in \mathbb{Z}\}$. Its fundamental domain is compact and one can apply the standard variational methods to prove that $\mathfrak{M}_\alpha^{per} \neq \emptyset$ for $\alpha \in \mathbb{Q}^n$.

Here is a difficulty to be overcome: One has to show that a function u minimizing (1.7) over a fundamental domain with the above periodicity condition automatically minimizes (1.7) in the sense of (1.11) in the large. Here the scalar nature (maximum principle) is essential (see [17]). This way we obtain $\mathfrak{M}_\alpha \neq \emptyset$ for $\alpha \in \mathbb{Q}^n$.

If $\alpha \in \mathbb{R}^n$ is arbitrary we pick an approximating sequence $\rho_m \in \mathbb{Q}^n$; $\rho_m \to \alpha$ and a sequence $u_m \in \mathfrak{M}_{\rho_m}$. From Theorem 4.6 we conclude

Theorem 4.7: $\mathfrak{M}_\alpha \neq \emptyset$ for all $\alpha \in \mathbb{R}^n$.

Incidentally, $\mathfrak{M}_\alpha^{per}$ is generally a proper subset of $\mathfrak{M}_\alpha$ for rational α. The nonperiodic $u \in \mathfrak{M}_\alpha$, $\alpha \in \mathbb{Q}^n$ have recently been investigated by V. Bangert. [V. Bangert, lecture presented at Oberwolfach meeting on Dynamical Systems, May 1987]

5. Group Action of $\mathbb{Z}^{n+1}$

a) In the following we will assume that α is not a rational vector and investigate the group action of $\mathbb{Z}^{n+1}$ on $\mathfrak{M}_\alpha$ given by

$$u \to \tau_{\bar{j}}u \quad \text{for} \quad \bar{j} \in \mathbb{Z}^{n+1} .$$

If $u \in \mathfrak{M}$ so is the whole orbit, as well as its closure, denoted by $\mathfrak{M}(u)$, in $\mathfrak{M}_\alpha$. Thus $\mathfrak{M}(u) \subset \mathfrak{M}_\alpha$ is a closed invariant set under this group action. Moreover, by Proposition 4.4, $\mathfrak{M}(u)$ is totally ordered, i.e. for $v,w \in \mathfrak{M}(u)$ we have either $v < w$ or $v > w$ or $v \equiv w$, and this order is preserved by the action. Therefore one gets a picture of the situation by looking at the closed set

$$S_o = \{v(0) \mid v \in \mathfrak{M}(u)\}$$

on the real axis $\mathbb{R}$, or on the circle $\mathbb{R}/\mathbb{Z}$. This closed set is invariant under the commuting mappings f^k defined in Section 3. For such mappings there exists a unique minimal set, where "minimal" is to be understood in the sense of dynamical systems: A nonempty closed invariant set without proper subset with these properties. This minimal set consists of the set of cluster points of S_o, i.e. of points which are limit points of decreasing or increasing sequences of S_o. This minimal set is a Cantor set on $\mathbb{R}$, frequently called Denjoy set, unless it agrees with $\mathbb{R}$.

Similarly, the orbit closure $\mathfrak{M}(u)$ of $u \in \mathfrak{M}_\alpha$ contains a unique minimal set consisting of functions $v \in \mathfrak{M}(u)$ which are limits (say in the C^o-topology) of subsequences of the translates $\tau_{\bar{j}}v$ with $(\bar{j},\bar{\alpha}) \neq 0$. In dynamical systems such orbits are called <u>recurrent</u> and therefore we will denote this set by $\mathfrak{M}^{rec}(u)$. It can be characterized as follows:

$\mathfrak{M}^{rec}(u)$ consists of all $v \in \mathfrak{M}(u)$ which are limits of subsequences of the orbit $\{\tau_{\bar{j}}v\}$, $(\bar{j},\bar{\alpha}) \neq 0$

or equivalently, $\mathfrak{M}^{rec}(u)$ is the unique minimal closed invariant subset of $\mathfrak{M}(u)$. (see Bangert [3]). Thus

$$S_o^{rec} = \{v(0) \mid v \in \mathfrak{M}^{rec}(u)\} \tag{5.1}$$

is either the whole real axis, or a Cantor set obtained from S_o by removing the isolated points.

b) Recently Bangert [3] proved the basic fact, that this set $\mathfrak{M}^{rec}(u)$ is independent of the choice of $u \in \mathfrak{M}_\alpha$. This fact is, of course, fundamental for this theory because it associates with each $\alpha \in \mathbb{R}^n \setminus \mathbb{Q}^n$ a distinguished set of minimals which therefore has geometric significance. It is the analogue of the Aubry-Mather set for monotone twist sets:

<u>Theorem 5.1</u> (V. Bangert): If $\alpha \in \mathbb{R}^n \setminus \mathbb{Q}^n$ and $u,v \in \mathfrak{M}_\alpha$ then

$$\mathfrak{M}^{rec}(u) = \mathfrak{M}^{rec}(v) .$$

For the interesting proof we refer to Bangert [3].

Therefore we will denote this set by $\mathfrak{M}_\alpha^{rec}$. Since it is a minimal closed invariant set of $\mathfrak{M}(u)$ under the group action one can characterize $\mathfrak{M}_\alpha^{rec}$ also as a minimal closed invariant set of $\mathfrak{M}_\alpha$ and it follows:

Corollary 5.2: For $\alpha \in \mathbb{R}^n \setminus \mathbb{Q}^n$ the set $\mathfrak{M}_\alpha$ has a unique minimal closed invariant subset $\mathfrak{M}_\alpha^{rec}$.

For rational α it is reasonable and consistent to define

$$\mathfrak{M}_\alpha^{rec} = \mathfrak{M}_\alpha^{per} \quad \text{for} \quad \alpha \in \mathbb{Q}^n$$

(see (4.9)). For the case $n = 1$, $\mathfrak{M}_\alpha^{per}$ consists of the set of periodic orbits of minimal period.

We remark that in the generic case, when $\alpha_1, \alpha_2, \ldots, \alpha_n, -1$ are rationally independent the Theorem 5.1 implies that for given $a \in \mathbb{R}$ there exists at most one $u \in \mathfrak{M}_\alpha$ with $u(0) = a$. (see Bangert [3]). Thus we have the unusual situation that a minimal is uniquely determined by the asymptotic behaviour, given by α and the value at a single point, given by a. This is reminiscent of Liouville's theorem for harmonic functions.

It is clear that the elements of $\mathfrak{M}_\alpha^{rec}$ can be approximated in the C^0 topology by periodic minimals. Indeed, if α is not rational, then some $u \in \mathfrak{M}_\alpha$ can be so approximated; this was the basis of the proof of theorem 4.7. Therefore every translate of $\tau_{\bar{j}} u$ can be so approximated, hence every element of the orbit closure $\mathfrak{M}(u)$ which contains $\mathfrak{M}_\alpha^{rec}$. Hence

$$\text{closure}\left(\bigcup_{\rho \in \mathbb{Q}^n} \mathfrak{M}_\rho^{per} \right) \supset \bigcup_\alpha (\mathfrak{M}_\alpha^{rec}) := \mathfrak{M}^{rec} . \tag{5.2}$$

One could conjecture that both sides are equal, i.e. that the set of recurrent minimals is the closure of periodic minimals, but this has not been proven.

This statement corresponds to the analogue statement for monotone twist maps that invariant curves, or Mather sets, lie in the closure of minimal periodic orbits.

c) Analytic description of $\mathfrak{M}(u)$. We assume that α is not a rational vector and let $u \in \mathfrak{M}_\alpha$. Then the orbit $\{\tau_{\bar{j}} u\}$ is totally ordered. As a matter of fact, from (4.3) we conclude that

$$\tau_{\bar{j}} u_o > u_o \Rightarrow \tau_{\bar{j}} u > u$$

where $u_o = (\alpha, x)$. (However, $\tau_{\bar{j}} u_o = u_o$ does not always imply $\tau_{\bar{j}} u = u$!). For simplicity of the discussion we assume that the $\alpha_1, \alpha_2, \ldots, \alpha_n, -1$ are rationally independent, so that $\tau_{\bar{j}} u_o = u_o$ occurs only for $\bar{j} = 0$. In that case the above relation shows that the relation $(\bar{j}, \bar{\alpha}) = (j, \alpha) - j_{n+1} \to u(x+j) - j_{n+1}$ is monotone. We define the monotone function $U(x, \theta)$ by the conditions

$$U(x, \theta) = u(x+j) - j_{n+1} \quad \text{if} \quad \theta = (\alpha, x) + (\bar{j}, \bar{\alpha})$$

or

$$U(x, \tau_{\bar{j}} u_o) = \tau_{\bar{j}} u .$$

For every $x \in \mathbb{R}^n$ the function U is defined on a dense set and is strictly monotone in θ. Moreover, we have

$$(5.3) \qquad \begin{cases} U(x+e_\nu,\theta) = U(x,\theta) , & (\nu = 1,\dots,n) \\ U(x,\theta+1) = U(x,\theta) + 1 . & \end{cases}$$

We can extend the domain of definition of U to all of $\mathbb{R}^{n+1}$ using monotonicity:

$$(5.4) \qquad \begin{cases} U^+(x,\theta) = \lim\limits_{\theta_s \downarrow \theta} U(x,\theta_s) \\ U^-(x,\theta) = \lim\limits_{\theta_s \uparrow \theta} U(x,\theta_s) \end{cases}$$

where θ_s corresponds to values where U is defined. Then, for every $x \in \mathbb{R}^n$

$$U^+ = U^-$$

for all but a denumerable number of θ, at which U^+ and U^- have a discontinuity. In any event $U^- \leq U \leq U^+$, and $U^+ = U^-$ almost everywhere. There are two cases

<u>A)</u> $U^+ = U^-$ for all x,θ, i.e. they define a continuous function $U = U(x,\theta)$ and the mapping

$$(5.5) \qquad (x,\theta) \to (x,U(x,\theta))$$

defines a homeomorphism of the torus T^{n+1} taking the foliation

$$\theta = (\alpha,x) + \beta$$

into the desired minimal foliation

$$x_{n+1} = U(x,\alpha x+\beta) .$$

In this case every leaf is dense on the torus and $\mathfrak{M}_\alpha^{rec} = \mathfrak{M}_\alpha$.

<u>B)</u> $U^+ \not\equiv U^-$, i.e. for fixed x the set of θ for which $U^-(x,\theta) < U^+(x,\theta)$ is non-empty; it is a denumerable set. For example for $x = 0$ the set is the union of the open intervals

$$(5.6) \qquad \bigcup_m (U^-(0,\theta_m), U^+(0,\theta_m))$$

where θ_m are all discontinuities of U^+, U^-. It is the complement of a Cantor set. The minimals passing through this Cantor set are given by

$$(5.7) \qquad u(x) = U^{\pm}(x,(\alpha,x)+\theta) , \qquad \theta \in \mathbb{R} .$$

Again all these minimals are obviously recurrent, and because of the theorem 5.1 by Bangert all elements of $\mathfrak{M}_\alpha^{rec}$ are given by (5.7) and the complement of (5.6) is the set S_0^{rec} given by (5.1).

In case A) all minimals graph u, $u \in \mathfrak{M}_\alpha = \mathfrak{M}_\alpha^{rec}$ are dense on the torus and define a foliation. In the case B) the graph u is not dense on the torus for any

$u \in \mathfrak{M}_\alpha^{rec}$; they form the leaves of a lamination which can be visualized as a family of leaves covering only the part of the torus, namely the complement of the set of gaps

$$\bigcup_m \{\bar{x} \in \mathbb{R}^{n+1}, U^-(x,(\alpha,x)+\theta_m) < x_{n+1} < U^+(x,(\alpha,x)+\theta_m)\} \tag{5.8}$$

We remark the width of such a gap

$$d_m(x) = u_m^+(x) - u_m^-(x) \quad ; \quad u_m^{\pm}(x) = U^{\pm}(x,(\alpha,x)+\theta_m)$$

is uniformly wide over any compact domain, i.e.

$$c^{-1}d_m(y) \leq d_m(x) \leq cd_m(y) \quad \text{for} \quad x,y \in B_R$$

where $c \geq 1$ depends on R but not on m. This is a consequence of the Harnack inequality for functions of the de Giorgi class, as derived by di Benedetto and Trudinger [4].

All these statements, described here for rationally independent $\alpha_1,\alpha_2,\ldots,\alpha_n,-1$ are valid for $\alpha \in \mathbb{R}^n \setminus \mathbb{Q}^n$, i.e. not rational vectors α. The case of rational α which always involves compact leaves will not be discussed here.

d) In case A) the foliation is described by a continuous function $U = U(x,\theta)$ satisfying the periodicity conditions (5.3) and which is strictly monotone in θ. Moreover, since

$$u(x) = U(x,(\alpha,x)+\text{const})$$

are minimals it follows that $U(x,\theta)$ satisfies the differential equation

$$\sum_{\nu=1} D_\nu F_{p_\nu}(x,U,DU) = F_u(x,U,DU) \tag{5.9}$$

$$D_\nu = \frac{\partial}{\partial x_\nu} + \alpha_\nu \frac{\partial}{\partial \theta}$$

conversely, any solution of this differential equation with the above periodicity and monotonicity condition gives rise to a minimal foliation corresponding to $\alpha \in \mathbb{R}^n$.

In case B) the two functions U^+, U^- which differ only on a set of measure zero satisfy the same condition, except they are not continuous in θ. If we require upper semicontinuity in θ, for definiteness, we single out the function U^+.

<u>Theorem 5.3</u>: Given $\alpha \in \mathbb{R}^n \setminus \mathbb{Q}^n$ there exists a function $U = U(x,\theta)$ in $\mathbb{R}^{n+1}$, strictly monotone, and upper semicontinuous in θ for fixed x, such that $U(x,(\alpha,x)+\text{const})$ is twice continuously differentiable and satisfies (5.9) and the periodicity condition (5.3) such that all but denumerable many $u \in \mathfrak{M}_\alpha^{rec}$ are represented in the form

$$u(x) = U(x,(\alpha,x)+\text{const}) \quad ,$$

From our construction and theorem 5.1 it is clear that U is a.e. uniquely determined. The case A) corresponds to continuous U and the B) to discontinuous U. In this case the mapping (5.5) is not a homeomorphism of $\mathbb{R}^{n+1}$ onto itself but of $\mathbb{R}^{n+1}$ onto the complement of the set (5.8).

Theorem 5.1 shows that our problem, to determine $\mathfrak{M}_\alpha^{rec}$ is reduced to finding a weak solution U to the above partial differential equation (5.9) with periodicity and monotonicity condition. An alternate way to determine such a solution U and hence $\mathfrak{M}_\alpha^{rec}$ will be described in lecture IV.

e) Description of the foliation in terms of a one-form. In case A) the continuous function $U = U(x,\theta) = x_{n+1}$ has an inverse function

$$\theta = Z(\bar{x})$$

monotone in x_{n+1}, such that $Z(\bar{x}) - x_{n+1}$ has period 1 in all variables. Thus the foliation in question is given as level sets of the function

$$z(\bar{x}) = (\alpha,x) - Z(\bar{x})$$

or, in terms of the closed one form (in the sense of distributions)

$$\omega = dz = \sum_{\nu=1}^{n} \alpha_\nu dx_\nu - dZ \ .$$

Its periods over the circles $\gamma_\nu = \{\bar{x} = te_\nu, \ 0 \le t \le 1\}$ on T^{n+1} are given by

$$\int_{\gamma_\nu} \omega = z(\bar{x}+e_\nu) - z(\bar{x}) = \alpha_\nu \qquad (\nu = 1,\dots,n+1)$$

with $\alpha_{n+1} = -1$. Moreover, the function z is strictly monotone in x_{n+1} . This describes the foliation in the more customary form in terms of a one-form and exhibits its holomony. In case B) the function $z(\bar{x})$ will be constant on the set (5.8).

III) Preservation and Desintegration of a Smooth Foliation

6. A stability theorem

a) The question of interest is the following: which of the two cases A or B arises in a given situation. In particular, if a variational problem depends on a parameter, as for example in (2.3), what can be said about the existence of a smooth foliation in dependence on such a parameter. For the particular example given in Section 2,(2.3) it was proven by Bangert [2] for any positive number A there exists a bound $\lambda^*(A)$ such that for $\lambda > \lambda^*(A)$ there exists no continuous foliation for $|\alpha| \leq A$. In other words for all $\alpha \in \mathbb{R}^n$ with $|\alpha| \leq A$ the sets $\mathfrak{M}_\alpha^{rec}$ are Cantor sets, and the functions $U^+(x,\theta)$ constructed in Section 5 have discontinuities. On the other hand for $|\lambda|$ sufficiently small there are smooth minimal foliations for this example (2.3) as the following theorem guarantees. Incidentally, for any fixed λ one can construct smooth minimal foliations of certain α with $|\alpha|$ large enough. Thus the situation is rather subtle and a general answer can not be expected. However, one can give simple geometrical criteria which are sufficient for the non-existence of continuous foliations. First we discuss the opposite situation of smooth foliations.

b) We begin with a perturbation theorem of the type of theorem 1.1. For the formulation of the result we assume that the integrand $F = F(\bar{x},p)$ satisfies the periodicity condition i) and the Legendre condition ii) of (1.8). However, no growth condition (like (1.8) iii)) is needed but we require instead

$$F \in C^\infty(\Omega)$$

where Ω is an open domain in $(T^{n+1} \times \mathbb{R}^n)$ with $\pi_1 \Omega = T^{n+1}$, where $\pi_1(\bar{x},p) = \bar{x}$. The projection of Ω on the p-direction is assumed to be bounded. Moreover, we assume that Ω is invariant under the translations $(\bar{x},p) \to (\bar{x}+\bar{j},p)$ for all $\bar{j} \in \mathbb{Z}^{n+1}$.

We assume that $F = F^\lambda \in C^\infty(\Omega)$, depends continuously on a real parameter $\lambda \in (-\delta,+\delta)$, e.g.

$$F^\lambda = F^0 + \lambda G .$$

Thus if the Legendre condition holds for $\lambda = 0$ then also for sufficiently small $|\lambda|$.

We stipulate that for $F = F^0$ and a given rotation vector α there exists a smooth foliation which is given by a smooth function $U = U^0(x,\theta)$, such that

$$(x,U^0(x,\theta),DU^0) \in \Omega \quad \text{for all} \quad (x,\theta) \in T^{n+1}$$

and such that U^0 satisfies the conditions (5.3), (5.9), with F replaced by F^0, and the monotonicity condition in the strict form

$$\frac{\partial U^o}{\partial \theta} > 0 . \tag{6.2}$$

We ask for a smooth F^λ-minimal foliation for the <u>same</u> rotation vector α, provided $|\lambda|$ is sufficiently small. As was explained in Section 1 this requires a Diophantine condition on α which we put in the following form:

There should exist positive numbers γ, τ such that for all $\bar{j} = (j, j_{n+1}) \in \mathbb{Z}^{n+1} \setminus (0)$ the inequalities

$$(6.3) \qquad \sum_{\nu=1}^{n} (\alpha_\nu j_{n+1} + j_\nu)^2 \geq \gamma^{-1}(1+j_{n+1}^2)^{-\tau}$$

hold. (The relationship between (6.3) and (1.5) will be explained below).

<u>Theorem 6.1</u>: Under the above hypotheses there exist, for sufficiently small $|\lambda|$, a smooth solution $U = U^\lambda(x,\theta)$ of the partial differential equation (5.9), with F replaced by F^λ, satisfying the periodicity condition (5.3). Moreover, U^λ depends continuously on λ in the $C^\infty(T^{n+1})$-topology and $U^\lambda = U^0$ for $\lambda = 0$; in particular, from (6.2)

$$(6.4) \qquad \frac{\partial U^\lambda}{\partial \theta} > 0 .$$

The proof of this theorem uses a rapidly convergent iteration technique as it is applied in the so-called K.A.M. theory. As a matter of fact, this theorem is a generalization of the theorem on the existence of invariant tori for Hamilton systems of two degrees of freedom to partial differential equation. While the traditional proof of invariant tori makes use of canonical transformation theory this is not possible for partial differential equation, since the analogues of canonical transformation are trivial. Therefore we have to work in the "configuration space" and apply the implicit function theorem directly to (5.9). This approach gives rise to a simplified proof of the existence of invariant to tori for n degrees of freedom, as was shown recently by Salamon and Zehnder [19].

c) According to Theorem 6.1 the λ-values for which smooth minimal foliations exist with a fixed rotation vector α satisfying (6.3) form an open set. These foliations which we denote by $\mathcal{F}^\lambda = \mathcal{F}^\lambda_\alpha$ are conjugate to $\mathcal{F}^0$. Indeed, if $\mathcal{F}^\lambda$ is given by

$$x_{n+1} = U^\lambda(x,\theta) \ ; \ \theta = (\alpha, x) + \text{const.}$$

then the mapping

$$\phi^\lambda : (x,\theta) \to (x, U^\lambda(x,\theta))$$

defines a diffeomorphism of T^{n+1} onto itself and the diffeomorphism

$$\psi^\lambda = \phi^\lambda \circ (\phi^0)^{-1}$$

takes $\mathcal{F}^0$ into $\mathcal{F}^\lambda$.

Theorem 6.1 can be given a stronger form: Instead of considering one parameter families of integrands one can consider a $C^\infty(T^{n+1} \times \Omega)$ neighborhood of $F^0 = F^0(\bar{x},p)$.

In other words, one can prove

Theorem 6.2: Under the hypothesis of theorem 6.1 there exists a $C^\infty(T^{n+1}\times\Omega)$-neighborhood $\mathbf{N} = \mathbf{N}_\alpha(F^o)$ such that for every $F\in\mathbf{N}$ there exists a solution $U = U(x,\theta)$ of (5.9) and (5.3), depending continuously on F.

The foliations $\mathcal{F}_U: x_{n+1} = U(x,\theta)$ are again equivalent to $\mathcal{F}^o: x_{n+1} = U(x,\theta)$ under the diffeomorphism

$$\psi = \phi_U\circ(\phi^o)^{-1}$$

where

(6.5)
$$\phi_U: (x,\theta)\longmapsto(x,U(x,\theta)) .$$

Hence, these foliations are "stable" if we define a smooth foliation $\mathcal{F}^o$, which is minimal with respect to F^o, stable, if for all F in a C^∞-neighborhood $\mathbf{N}$ of F^o such that for all $F\subset\mathbf{N}$ there exists a smooth F-minimal foliation $\mathcal{F}$ conjugate to $\mathcal{F}^o$. Therefore we have

Corollary 6.3: Under the hypotheses of Theorem 6.1 the foliation $\mathcal{F}^o$ is stable.

d) We apply these results to the minimal foliation discussed in Section 1 b). Theorem 6.2 implies Theorem 1.1 although the stability concept of Section 1 is different from the one used here since it refers to the narrower class of integrands of the form

$$F(\bar{x},p) = D(\bar{x})\sqrt{\sum_{\nu,\mu=1}^{n+1} g^{\nu\mu}(\bar{x})p_\nu p_\mu} \quad \text{where } p_{n+1} = -1 ,\ D = \sqrt{\det(g_{\nu\mu})} .$$

In particular, we have (in the example of Section 1, b))

$$F^o = \sqrt{1+|p|^2} .$$

Note that these integrands violate the quadratic growth condition (1.8) iii) but Theorem 6.2 requires only for F to be defined in an open domain Ω containing the foliation $\mathcal{F}^o$. Here we can take

$$\Omega = \{(\bar{x},p)\in T^{n+1}\times\mathbf{R}^n ,\quad |p-\alpha|<\delta\} ,$$

and a growth condition is irrelevant.

We remark that the Diophantine condition (6.3) actually is equivalent with (1.5) for a vector $\bar{\alpha}$ with $\alpha_{n+1} = -1$. More precisely, the condition (1.5) with some constant $c_o\geq 1$ implies

(6.6)
$$\delta(\bar{j}) := \sum_{\nu=1}^{n}(\alpha_\nu j_{n+1}-j_\nu)^2 \geq \gamma^{-1}(1+j_{n+1}^2)^{-\tau}$$

for all $\bar{j}\in\mathbf{Z}^{n+1}\setminus(0)$ if $\gamma = (2|\bar{\alpha}|^2)^\tau c_o$.

Conversely, this latter condition with $\gamma = (2^{\tau}|\alpha|)^{-1}c_0$ implies (1.5).

The proof follows from the inequality

$$\delta(\bar{x}) \leq \sum_{1\leq\nu<\mu\leq n+1} (\alpha_\nu x_\mu - \alpha_\mu x_\nu)^2 = |\bar{\alpha}|^2|\bar{x}|^2 - (\bar{\alpha},\bar{x})^2 \leq |\bar{\alpha}|^2\delta(\bar{x})$$

for all $\bar{x} \in \mathbf{R}^{n+1}$ which is easily verified. Therefore (6.6) implies (1.5) if we observe that

$$|\bar{j}|^2 = |j|^2 + j_{n+1}^2 \geq \frac{1}{2}(1+j_{n+1}^2) .$$

On the other hand, (1.5) implies

$$\delta(\bar{j}) \geq c_0^{-1}|\bar{j}|^{-2\tau} .$$

Now we observe, that by the triangle inequality

$$\delta(\bar{j}) \geq \frac{1}{2}|j|^2 - |\alpha|^2 j_{n+1}^2$$

hence $\delta(\bar{j}) \geq \frac{1}{2}$ if $|j|^2 \geq 2|\alpha|^2 j_{n+1}^2 + 1$, proving the inequality in this case with $\gamma \geq 2$. But in the remaining case: $|j|^2 \leq 2|\alpha|^2 j_{n+1}^2 + 1$ we have

$$|\bar{j}|^2 \leq (2|\alpha|^2+1)j_{n+1}^2 + 1 \leq 2|\bar{\alpha}|^2(j_{n+1}^2+1)$$

hence

$$\delta(\bar{j}) \geq c_0^{-1}(2|\bar{\alpha}|^2(1+j_{n+1}^2))^{-\tau} ,$$

as we wanted to show.

e) The assumptions underlying Theorem 6.1 are very restrictive: α) the Diophantine condition (6.3), β) the smallness condition on $|\lambda|$. We want to point out that both these conditions are necessary.

Mather proved in [14] a result for monotone twist mappings which can be translated into the following statement for minimal foliations on T^2, i.e. for $n = 1$. In that case the condition requires constants γ, τ such that

$$|q\alpha - p| \geq \gamma^{-1} q^{-\tau}$$

for all rational numbers p/q. A number α for which such constants do not exist is called a Liouville number. They are characterized by the property that for any large numbers γ, τ there exists a rational number p/q so that

$$0 < |q\alpha - p| < \gamma^{-1} q^{-\tau} .$$

An example of such a Liouville number is

$$\alpha = \sum_{\nu=1}^{\infty} 2^{-(\nu!)} .$$

The set of Liouville numbers form a set of measure 0.

Mather's result takes the following form: If $F^o \in C^\infty(\Omega)$ satisfies the above hypotheses and possesses a smooth foliation $\mathcal{F}^o$ (satisfying also (6.3)) with rotation number α which is a Liouville number then any C^∞ neighborhood $\mathbf{N} = \mathbf{N}(F^o)$ contains a $F \subset \mathbf{N}$ not admitting a smooth foliation with respect to this α .

In other words, the Diophantine condition is necessary for Theorem 6.2. As for the smallness condition for $|\lambda|$ we mentioned already in Section 2 the example (2.3) for which Bangert showed the nonexistence of smooth minimal foliation for $|\lambda| > \lambda^*(A)$ of rotation vector α , $|\alpha| \leq A$.

Thus in this example we have a smooth minimal foliation $\mathcal{F}^\lambda(\alpha)$ for $|\lambda|$ sufficiently small, always assuming (6.3), but for some critical value of λ the foliation disintegrates to a "lamination", developing gaps densely on the torus. Analytically this means that the solution $U = U^\lambda(x,\theta)$ describing the $\mathcal{F}^\lambda(\alpha)$ is smooth for small values of $|\lambda|$ but becomes discontinuous beyond some critical value.

f) The existence of smooth foliations may, more trivially, be due to symmetries of the variational problem. If, for example, $\frac{\partial}{\partial x_{n+1}} F(\bar{x},p) \equiv 0$, then the variational problem is invariant under the translation $x_{n+1} \to x_{n+1} + \text{const}$ and, if $u(x)$ is any minimal then so is $u(x) + \text{const}$ minimal. The foliation is then defined by

$$U(x,\theta) = \theta + u(x)$$

which is obviously continuous.

Similarly, if $\frac{\partial}{\partial x_1} F \equiv 0$, then with $u(x)$ is $u(x+se_1)$ minimal. To show that $x_{n+1} = u(x+se_1)$ defines a foliation it suffices to prove the following

<u>Proposition 6.4</u>: If $u \in \mathfrak{M}_\alpha$ and

$$\frac{\partial F}{\partial x_1} \equiv 0 , \quad \alpha_1 \neq 0$$

then

$$\alpha_1 \frac{\partial u}{\partial x_1} > 0 .$$

<u>Proof</u>: We may assume that

$$\alpha_1 > 0 .$$

If

$$\min \frac{\partial u}{\partial x_1} < 0$$

we conclude from

$$\text{(6.7)} \qquad \frac{u(x+se_1)-u(x)}{s} \begin{cases} \to \alpha_1 & \text{for } s \to \infty \\ \to \frac{\partial u}{\partial x_1} & \text{for } s \to 0 \end{cases}$$

that there exists an s^* so that

$$f(s) = \min_x (u(x+se_1) - u(x)) = 0 \quad \text{for} \quad s = s^* .$$

The maximum principle (Prop. 4.4) implies that $u(x+s^*e_1) \equiv u(x)$ which contradicts (6.7). Hence $u_{x_1} \geq 0$, and again by the maximum principle we conclude $u_{x_1} > 0$.

These statements can be generalized:

Theorem 6.5: If $\bar{\gamma} \in \mathbb{R}^{n+1}$ and

$$F(\bar{x}+s\bar{\gamma},p) \equiv F(\bar{x},p) \quad \text{for all} \quad s \in \mathbb{R}$$

and

$$(\bar{\alpha},\bar{\gamma}) \neq 0 \ , \quad \bar{\alpha} = (\alpha,-1)$$

then any $u \in \mathfrak{M}_\alpha$ gives rise to a smooth minimal foliation given by

$$x_{n+1} = u(x+s\gamma) - s\gamma_{n+1} \ , \quad s \in \mathbb{R} .$$

In this theorem one will assume $\bar{\gamma}$ to be a rational vector since otherwise F must be independent of $\bar{x}$, i.e. $F = F(p)$. In this case the only minimal foliations are given by parallel hyperplanes.

7. A Mechanical Problem

a) In this Section we discuss the special case of $n = 1$ where

(6.8) $$F(t,x,\dot{x}) = \frac{1}{2}\dot{x}^2 + \lambda V(t,x) \ , \quad V \in C^\infty(T^2) .$$

Here we used the notation of mechanics, replacing x,u by t,x . The Euler equation is given by

(6.9) $$\ddot{x} = \lambda V_x(t,x) .$$

By Theorem 6.1 one has for any Diophantine number α a smooth minimal foliation of rotation number α if $|\lambda|$ is small enough. In other words, there exists a smooth solution U of the differential equation

(6.10) $$(\partial_t + \alpha\partial_\theta)^2 U = \lambda V_x(t,U)$$

with $U(t,\theta) - \theta$ of period 1 in both variables and $U_\theta > 0$.

As was discussed in Section 3 such a foliation gives rise to an invariant torus

(6.11) $$x = U(t,\theta) \quad , \quad \dot{x} = p = (\partial_t + \alpha\partial_\theta)U(t,\theta)$$

in the 3-dimensional phase space. Because of the importance of these tori for stability theory numerical studies have been undertaken to compute such invariant tori and to get realistic estimates for the critical λ-value, where such a torus disintegrates. We describe briefly the result of one such effort.

b) In the numerical work Escande (reference cited in [5] and by Celletti and Chierchia [5]) the potential function

$$V(t,x) = \frac{1}{2\pi}\{\cos(2\pi x) + \cos(2\pi(x-t))\}$$

was studied. One expects that the invariant tori will survive for relatively large λ if the rotation number α can not be well approximated by rationals. The standard example of this type is given by

$$\alpha = \frac{1}{2}(\sqrt{5}-1)$$

whose continued fraction expansion has only ones. The work of Escande gives numerical evidence, by following the orbits, that for $\lambda \sim \lambda_* = 0.169$ the invariant tori have disintegrated. On the other hand Celletti and Chierchia implemented the construction of invariant tori and showed the existence of smooth tori with rotation number α for

$$|\lambda| \leq 0.069 .$$

These two numbers are still quite a bit apart but one has to keep in mind that earlier attempts gave a discrepancy of some 10 orders of magnitude between the λ values for which the existence of invariant tori can be established and for which breakdown is observed.

c) We return to the question of dynamic stability for (6.9), discussed in Section 3. We asked whether for any solution of (6.9) the derivative $|\dot{x}(t)|$ can be unbounded. This is, in fact, not the case and in order to show that every solution has bounded derivatives it suffices to construct invariant tori for arbitrary large rotation numbers. We may assume $\lambda = 1$ and may achieve, by adding an irrelevant function of t alone that

$$\int_0^1 V(t,x)\,dx = 0 ,$$

so that there exists a periodic function $W \in C^\infty(T^2)$ with $V = W_x(t,x)$.

It turns out that the function

$$U^0(t,\theta) = \theta + \alpha^{-2} W(t,\theta)$$

is an approximate solution of our equation (6.10) for $\lambda = 1$ and $|\alpha|$ large. This becomes clear if one replaces the differential operator $\partial_t + \alpha\partial_\theta$ by $\alpha\partial_\theta$ and notes

$$\alpha^2 \partial_\theta^2 U^0 = V_x(t,\theta) .$$

If one also chooses α so as to satisfy a Diophantine condition, e.g. $\alpha_m = \alpha_0 + m$, $m \in \mathbb{Z}$ with α_0 a fixed such number, say $\frac{1}{2}(\sqrt{5}-1)$, then one can conclude that the differential equation (6.10) has a smooth solution for m large enough. (I want to thank L. Chierchia for discussions of this point.)

Such a solution gives rise to an invariant torus of the form (6.11). Eliminating θ we obtain an invariant torus in the form

$$\dot{x} = \psi(t,x)$$

where $\psi = \alpha_m + O(1) \to \infty$ as $m \to \infty$.

These tori provide bounds for $\dot{x}$ for all $t \in \mathbb{R}$ for any arbitrary solution $x = x(t)$ of (6.9): Given such a solution, choose α_m so large that $\dot{x}(0) \le \min_{t,x} \psi(t,x)$. Then it follows that $\dot{x}(t) \le \max_{t,x} \psi(t,x)$ for t .

d) For integrands of the form (6.8) it is rather easy to construct potentials V, such that for $|\lambda| \ge \lambda^*(A)$ there are no smooth invariant foliations of rotation number α with $|\alpha| \le A$, (see [18]). One simply constructs potentials V with support in a small neighborhood, containing a ball where $V \ge 1$, hence $\lambda V \ge \lambda$ is large for large λ . Therefore a minimal will not go through such a ball, since one can decrease the variational functional by going around this obstacle. The same idea underlies the counter example of Bangert [2], which requires, however, nontrivial estimates independent of λ for minimals in a compact region outside the support of V .

We point out that these examples assure for any fixed $\lambda \ge \lambda^*(A)$ the destruction of all smooth foliations only for $|\alpha| \le A$ and not for all α . The above remark c) shows that for these examples the latter can not be expected since for any given λ there are smooth minimal foliations for sufficiently large α . This is in sharp contrast to the difference equation

(6.12) $$x_{i+1} - 2x_i + x_{i-1} = \lambda V_x(x_i)$$

where $V = V(x)$ has period 1. For example, if $V(x) = \frac{-1}{2\pi}\cos 2\pi x$ one can show (see [22]) that the difference equation

$$U(\theta+\alpha) - 2U(\theta) + U(\theta-\alpha) = \lambda \sin(2\pi\, U(\theta))$$

has a monotone solution $U(\theta)$, satisfying $U(\theta+1) = U(\theta) + 1$ but which has discontinuities for $\lambda \ge \lambda^* = \frac{2}{3\pi}$ no matter what the value of α is. This difference equation is related to a particular monotone twist mapping, the so-called standard mapping. This remark shows: although the difference equation (6.12) has similar features as the differential equation (6.9) they exhibit different behavior for large frequencies α .

IV An Alternate Approach

8. The Regularized Variational Problem

a) In Section 5 we saw that any minimal foliation whose leaves are graphs can be described by a function $U = U(x,\theta)$ satisfying the partial differential equation (5.9) for some vector $\alpha \in \mathbb{R}^n$ the periodicity condition (5.3) and the monotonicity condition in θ. Now we consider the determination of such on U as an analytic problem removed from its connection with foliations and describe a straightforward construction of U based on a regularized variational problem.

Note that the equations (5.9) can be viewed as the Euler equations of

$$(8.1) \qquad \int_{\bar{Q}} F(x,U,DU)\,dx\,d\theta$$

where $\bar{Q} = [0,1]^{n+1}$ and $D_\nu = \partial_{x_\nu} + \alpha_\nu \partial_\theta$. This is an integral over an (n+1)-dimensional cube, and although F satisfy the Legendre condition for the variational problem (1.7) it violates the Legendre condition for (8.1). This problem is therefore degenerate. A second difficulty connected with (8.1) is that we require the extremal to be monotone in θ ; this gives difficulties in deriving the Euler equation for extremals. Both these difficulties are overcome by approximating (8.1) by a nondegenerate variational problem

$$(8.2) \qquad \int_{\bar{Q}} \{\frac{\varepsilon}{2} U_\theta^2 + F(x,U,DU)\}\,dx\,d\theta$$

which for $\varepsilon > 0$ has smooth minimals satisfying the periodicity conditions (5.3) and, in addition, they automatically are strictly monotone in θ. The desired solutions of (5.9) will be obtained by taking the limit $\varepsilon \downarrow 0$. In this limit the regularity may get lost, and again we arrive at the two cases A or B , depending whether the limit function is continuous or not. These cases can not be distinguished by this procedure, which is valid for arbitrary $\alpha \in \mathbb{R}^n$.

b) We discuss at first the existence and regularity theory for the problem (8.1) and set more generally

$$(8.3) \qquad J(U) = \int_{\bar{Q}} G(x,U,\bar{D}U)\,dx\,d\theta$$

where in the above example

$$G(\bar{x},\bar{p}) = \frac{\varepsilon}{2} p_{n+1}^2 + F(\bar{x},p)$$

$$D_\nu = \partial_{x_\nu} + \alpha_\nu \partial_\theta$$

$$D_{n+1} = -\partial_\theta \ , \quad \bar{D} = (D_1,D_2,\dots,D_{n+1})$$

However, we may allow, more generally, $G \in C^{2,\eta}(T^{n+1} \times \mathbb{R}^{n+1})$ satisfying the Legendre condition in $\bar{p}$ and a quadratic growth condition in $\bar{p}$, analogue to (1.8). For

$\varepsilon > 0$ the function G of (8.3) satisfies these conditions, though not uniformly for $\varepsilon \to 0$.

For $J(U)$ we consider the space of admissible functions

$$A = \{U \,|\, U(x,\theta) - \theta \in H^{1,2}(T^{n+1})\}$$

and minimize J over $U \in A$. Because of the quadratic growth condition J is defined in A and is lower semicontinuous on account of the assumed Legendre condition. Note $U^{0} = \theta$ belongs to A and

$$J(U^{0}) = \int_{\bar{Q}} G(x,0,\bar{\alpha})\,dx\,d\theta = c_1 ;$$

Thus $A \neq \emptyset$.

By standard arguments of the calculus of variation it follows that $\min J(U)$ is taken on in A.

For the following it will be essential that the functional J is invariant under the translation $U(x,\theta) \to U(x,\theta + \text{const})$. Especially since $U(x,\theta+1) - U(x,\theta) = 1$ we find that

$$\int_{\bar{Q}} (U(x,\theta+s) - U(x,\theta))\,dxd\theta = s$$

and we can normalize U so that

$$\int_{\bar{Q}} U(x,\theta)\,dxd\theta = 0 . \tag{8.4}$$

For a minimal so normalized we obtain an estimate for $\int_{\bar{Q}} (U^2 + |\nabla U|^2)\,dxd\theta$.

Using the technique of Giaquinta and Giusti again one can establish pointwise bounds and the regularity of a minimal, $U \in C^2$, satisfying the Euler equation

$$\sum_{\nu=1}^{n+1} D_\nu G_{p_\nu}(x,U,\bar{D}U) = G_{x_{n+1}}(x,U,\bar{D}U) \tag{8.5}$$

<u>Proposition 8.1</u>: Under the above hypotheses on G the functional $J(U)$ assumes its minimum in A at a function $U \in C^2$, satisfying the Euler equation. With $U = U(x,\theta)$ also $U(x,\theta+s)$ is a minimal.

<u>Proposition 8.2</u>: If U_1, U_2 are two C^2-solutions of the Euler equation (8.5) satisfying $U_1 \leq U_2$ in an open connected domain Ω then $U_1 < U_2$ in Ω or $U_1 \equiv U_2$ in Ω.

<u>Proposition 8.3</u>: If again $U_1, U_2 \in A$ are two C^2-solutions of the Euler equation (8.5) then there exists a constant s such that

$$U_2(x,\theta) = U_1(x,\theta+s)$$

i.e. these solutions in $A \cap C^2$ are unique up to translations.

<u>Proposition 8.4</u>: Any solution of the Euler equation $U \in A \cap C^2$ is strictly monotone and satisfies

$$\frac{\partial U}{\partial \theta} > 0 .$$

Proposition 8.1 was already discussed and the proof of Proposition 8.2 - 8.4 is a consequence of the maximum principle. Indeed, since $U_1 - U_2$ can be viewed as the solution of an elliptic partial differential equation it cannot attain a maximum in an open connected domain, proving Prop. 8.2.

To prove Prop. 8.3 set

$$f(s) = \min_{x,\theta}(U_1(x,\theta+s) - U_2(x,\theta)) .$$

Since $U_1(x,\theta+1) = U_1(x,\theta)+1$ for $U_1 \in A$ we note that $f(+\infty) = +\infty$, $f(-\infty) = -\infty$ hence since f is continuous, there exist an s^* with $f(s^*)=0$. Applying proposition 8.2 to $U_1(x,\theta+s^*) \geq U_2(x,\theta)$ gives proposition 8.3.

Finally to prove proposition 8.4 we consider $g(s) = \min_{x,\theta}(U(x,\theta+s) - U(x,\theta))$ and claim $g(s) \geq 0$ for $s \geq 0$, which would imply the monotonicity of U. Otherwise we would have $g(s) < 0$ for some positive s and since again $g(+\infty) = +\infty$ we would find a positive s^* with $g(s^*) = 0$. Applying proposition 8.2 to $U(x,\theta+s^*) \geq U(x,\theta)$ we would conclude

$$U(x,\theta+s^*) = U(x,\theta) \quad \text{for some} \quad s^* > 0 .$$

This would imply periodicity of U contradicting its unboundedness:

$$U(x,\theta+m) = U(x,\theta) + m \ , \ m \in \mathbb{Z} .$$

Hence U is monotone and hence $U_\theta(x,\theta) \geq 0$. Differentiating (8.5) with respect to θ we obtain an elliptic partial differential equation for U_θ and the maximum principle yields $U_\theta > 0$ unless $U_\theta \equiv 0$. The latter case is again excluded by the unboundedness of U.

These simple consequences of the maximum principle and the translation invariance of J under $\theta \to \theta + \text{const}$ has the following obvious implications: The minimal of J in A is unique up to translation. In fact, J has no other locally bounded extremals, i.e. there are no other weak solutions in A of (8.5) which are locally bounded. As a matter of fact, it is known (see [12], [9]) that all locally bounded extremals are regular, i.e. are C^2-solutions, but it is conceivable that (8.5) has weak solutions in A which are not locally bounded *). On the other hand all minimals are C^2 and the hardest part of the regularity proof is to establish that minimals are locally bounded.

---

*) For an example of this kind see J. Frehse, A note on the Hölder continuity of solutions of variational problems. Abhandlg. Math. Seminar, Univ. Hamburg <u>43</u>, 1975, pp 59–63. For this reference I am grateful to S. Hildebrandt.

Aside from these possible irregular extremals the functional J has no other extrema but the minimum in $\mathcal{A}$.

The monotonicity statement of prop. 8.4 shows that the minimals $U = U(x,\theta+s)$ form a field of extremals covering T^{n+2} in a simple manner.

c) We apply these results to the functional (8.2) which we denote by $J_\alpha^\varepsilon(U)$, the α-dependence entering through $D_\nu = \partial_{x_\nu} + \alpha_\nu \partial_\theta$. According to the above result there exist C^2-minimals in $\mathcal{A}$, denoted by $U = U_\alpha^\varepsilon$ so that

(8.6) $$M^\varepsilon(\alpha) = \min_{U\in\mathcal{A}} J_\alpha^\varepsilon(U) = J_\alpha^\varepsilon(U_\alpha^\varepsilon)$$

Moreover,

(8.7) $$\frac{\partial}{\partial\theta} U_\alpha^\varepsilon > 0 \,, \quad \int_0^1 \frac{\partial}{\partial\theta} U_\alpha^\varepsilon d\theta = 1 \,.$$

It is crucial to obtain estimates for $U = U_\alpha^\varepsilon$ which are independent of $\varepsilon > 0$. We find

<u>Proposition 8.5</u>: There exists a constant $c > 0$ independent of $\varepsilon > 0$, depending only on F and α such that for the minimal $U = U_\alpha^\varepsilon$ of J_α^ε one has

$$|D_\nu U|_{L^\infty} \,, \quad |D_\nu D_\mu U|_{L^\infty} \le c \quad \text{for} \quad \nu,\mu \le n$$

and

$$\sqrt{\varepsilon}\,|D_{n+1}U|_{L^\infty} \le c \,.$$

This statement shows that the derivatives along the leaves $\theta = (\alpha,x) + \text{const}$ can be bounded independent of ε while the derivative transversal to these leaves grows at most like $\varepsilon^{-1/2}$. The effect of the regularizing ε-term in (8.2) is to smoothen the discontinuities of U, similarly as in the theory of hyperbolic differential equations shocks are smoothed by a viscosity term.

The proof of proposition 8.5 can be obtained by rescaling the variational problem by the transformation:

$$\sigma_\varepsilon : (x,\lambda) \to (x, (\alpha,x) + \sqrt{\varepsilon}\lambda = \theta)$$

so that with $V = U\circ\sigma_\varepsilon$ we have

$$\int_{\bar{Q}} \frac{\varepsilon}{2} U_\theta^2 + F(x,U,DU)\,dx\,d\theta = \sqrt{\varepsilon} \int_{\sigma_\varepsilon^{-1}\bar{Q}} \frac{1}{2} V_\lambda^2 + F(x,V,V_x)\,dx\,d\lambda \,.$$

We can replace $\sigma_\varepsilon^{-1}\bar{Q}$ by another fundamental domain of $\mathbb{R}^{n+1}/\sigma_\varepsilon^{-1}\mathbb{Z}^{n+1}$, e.g.

$$P_\varepsilon = \left\{ x,\lambda \,\middle|\, |x_\nu| \le \frac{1}{2} \,,\ |\lambda| \le \frac{1}{2\sqrt{\varepsilon}} \right\}$$

and obtain for the minimal $U = U_\alpha^\varepsilon$ an estimate for

$$(8.8)\qquad \frac{1}{|P_\varepsilon|}\int_{P_\varepsilon}\{\frac{1}{2}V_\lambda^2 + F(x,V,V_x)\}dxd\lambda = \int_{\bar{Q}}\{\frac{\varepsilon}{2}U_\theta^2 + F(x,U,DU)\}dxd\theta \le c_1(\alpha)$$

by a constant independent of ε ; here $|P_\varepsilon| = \varepsilon^{-1/2}$ denotes the volume of P_ε . The same estimate holds, of course, for $2P_\varepsilon$ (of volume $2^{n+1}\varepsilon^{-1/2}$) in place of P_ε and with the assumptions on F we obtain

$$\frac{1}{|2P_\varepsilon|}\int_{2P_\varepsilon}(V_\lambda^2 + |V_x|^2)\,dxd\lambda \le c_2(\alpha)$$

independently of ε . We consider $2P_\varepsilon$ as the union of cubes

$$C_m = \{(x,\lambda) \in \mathbb{R}^{n+1},\ |x_\nu| \le 1,\ |\lambda-2m| \le 1\}$$

for integer m with $|m| \le N = [\frac{1}{\sqrt{\varepsilon}} - 1]$, so that for at least one of these cubes $C = C_{m^*}$ we have

$$\frac{1}{|C|}\int_C (V_\lambda^2 + |V_x|^2)\,dxd\lambda \le c_2(\alpha)\ .$$

By the de Giorgi-technique (see e.g. [12]) one obtains now for this rescaled variational problem

$$(8.9)\qquad \underset{\frac{1}{2}C}{\operatorname{osc}}\, V \le c_3(\alpha)$$

If one combines this with the fact that V is a monotone increasing function of λ and satisfies by (8.7)

$$\int_{-\varepsilon^{-1/2}}^{+\varepsilon^{-1/2}} \frac{\partial}{\partial\lambda} V\, d\lambda = 2$$

one derives easily from (8.9)

$$\underset{P_\varepsilon}{\operatorname{osc}}\, V \le c_4(\alpha)\ .$$

Now we apply the standard regularity theory [12] to the rescaled ε-independent variational problem (8.8) to obtain pointwise estimates for

$$|V_\lambda|, |V_x|, |V_{xx}| \le c_5(\alpha)$$

etc. in P_ε . This translates into the statement of the proposition. This argument was used by Denzler [6] in the case $n = 1$.

d) We assume that $U = U_\alpha^\varepsilon$ are the minimals of (8.2) normalized by (8.4). Then we can find a sequence $\varepsilon_m > 0$ tending to zero such that

$$U_\alpha^{\varepsilon_m} \to U_\alpha^0 \qquad \text{a.e.}$$

to a function satisfying the Euler equation (5.9), the periodicity and monotonicity condition. The proof depends on the above estimates and the weak form of the Euler equation for $U = U_\alpha^\varepsilon$:

$$\int_{\bar{Q}} \{\frac{\varepsilon}{2}\phi_\theta U_\theta + \sum_\nu F_{p_\nu}(x,U,DU)D_\nu\phi + F_u\phi\}dxd\theta = 0$$

for all $\phi \in C^1(T^{n+1})$. Since $\sqrt{\varepsilon}|U_\theta|$ is uniformly bounded one verifies easily that $U = U_\alpha^o$ satisfies

$$\int_{\bar{Q}} \{\sum_\nu F_{p_\nu}(x,U,DU)D_\nu\phi + F_u(x,U,dU)\phi\}dxd\theta = 0$$

for all $\phi \in C^1(T^{n+1})$, the weak form of (5.9). Thus $U_\alpha^o(x,(\alpha,x)+\theta) = u(x)$ is for almost all θ a solution of the original Euler equation (1.10). Of course, U_α^o will not be continuous in θ , in general. We forego the proof that U_α^o agrees almost everywhere with the function $U^{\pm}$ constructed in Section 5.

This concludes our sketchy discussion of the construction of the function $U^{\pm}$ with the aid of a regularized variational problem.

It is worth noting that the (n+1)-dimensional integral J_α^ε can be obtained from the n-dimensional integral by averaging:

$$J_\alpha^\varepsilon(U) = \lim_{r\to\infty}\frac{1}{|B_r|}\int_{B_r}(\frac{\varepsilon}{2}U_\theta^2 + F(x,U,DU))dx \quad \text{where} \quad \theta = (\alpha,x)+\beta$$

provided α is not rational, for any C^1-function. This is a consequence of the equidistribution of the foliation $x_{n+1} = (\alpha,x) + \text{const}$ if α is not rational.

e) In conclusion we want to show that the minimal energy $M^\varepsilon(\alpha)$ given by (8.6) is a convex function of α . We show this for $\varepsilon > 0$ when $M^\varepsilon(\alpha)$ is a smooth function of α and conclude the convexity of $M^o(\alpha)$ by a limit process. For $n = 1$ the convexity of the minimal average energy was discovered by Aubry and Mather [20]; it seems that their argument is not applicable for $n > 1$.

Differentiating (8.6) with respect to α_ν we obtain for $\varepsilon > 0$, using that the first variation of J_α^ε vanishes, the formula

$$\frac{\partial}{\partial\alpha_\nu}M^\varepsilon(\alpha) = \int_{\bar{Q}} F_{p_\nu}(x,U,DU)U_\theta dxd\theta$$

where U stands for the minimal U_α^ε . Note that for $\varepsilon > 0$ U_α^ε and $M^\varepsilon(\alpha)$ are twice continuously differentiable in α . If one defines the functions $\psi_\nu(\bar{x})$ implicitly by the equation

$$\psi_\nu(x,U(x,\theta)) = D_\nu U(x,\theta)$$

which by (8.7) is well defined, one obtains

$$\frac{\partial}{\partial\alpha_\nu}M^\varepsilon(\alpha) = \int_{\bar{Q}} F_{p_\nu}(x,x_{n+1},\psi)d\bar{x} \; . \tag{8.10}$$

The calculation of the second derivatives of $M^\varepsilon(\alpha)$ is more complicated but the formula (8.10) suggests it depends on F only via the positive definite Hessian F_{pp} . This is indeed the case and we give the result of the computation.

We set

$$D_\xi = \sum_{\nu=1}^{n} \xi_\nu \frac{\partial}{\partial \alpha_\nu}$$

and will compute

$$D_\xi^2 M^\varepsilon(\alpha) = \sum_{\nu,\mu=1}^{n} \xi_\nu \xi_\mu \frac{\partial^2 M^\varepsilon}{\partial \alpha_\nu \partial \alpha_\mu} .$$

With $U = U_\alpha^\varepsilon$ and

$$W = W(x,\theta) = U_\theta^{-1}(D_\xi U)$$

we find

$$D_\xi^2 M^\varepsilon(\alpha) = \int_{\bar Q} U_\theta^2 \{\frac{\varepsilon}{2}(D_{n+1}W)^2 + \sum_{\nu,\mu=1}^{n} F_{p_\nu p_\mu}(x,U,DU)(\xi_\nu + D_\nu W)(\xi_\mu + D_\mu W)\} dx d\theta$$

which shows the convexity of $M^\varepsilon(\alpha)$ for $\varepsilon > 0$.

In fact, one obtains a lower bound

$$(8.11) \qquad D_\xi^2 M^\varepsilon(\alpha) \geq \delta \int_{\bar Q} U_\theta^2 \sum_{\nu=1}^{n} (\xi_\nu + D_\nu W)^2 dx d\theta \geq \delta \left(\int_{\bar Q} (U_\theta)^{-2} dx d\theta \right)^{-1} |\xi|^2 .$$

This is a consequence of Schwarz inequality

$$\xi_\nu^2 = \left(\int_{\bar Q} (\xi_\nu + D_\nu W) dx d\theta \right)^2 \leq \int_{\bar Q} U_\theta^2 (\xi_\nu + D_\nu W)^2 dx d\theta \int_{\bar Q} U_\theta^{-2} dx d\theta .$$

Similarly, one can derive an upper estimate for $D_\xi^2 M^\varepsilon$. It follows from a different formula for this second derivative which is derived with the help of the differential equation for W . One finds

$$D_\xi^2 M^\varepsilon = \int_{\bar Q} U_\theta^2 \sum_{\nu,\mu=1}^{n} F_{p_\nu p_\mu} \xi_\nu \xi_\mu dx d\theta - \int U_\theta^2 \{\frac{\varepsilon}{2}(D_{n+1}W)^2 + \sum_{\nu,\mu=1}^{n} F_{p_\nu p_\mu} D_\nu W D_\mu W\} dx d\theta$$

which gives

$$(8.12) \qquad D_\xi^2 M^\varepsilon \leq c \int_{\bar Q} U_\theta^2 dx d\theta |\xi|^2 .$$

The estimate (8.12) shows clearly that upper bounds for U_θ give rise to upper bounds for $D_\xi^2 M^\varepsilon$.

The calculation of the second derivative of M^ε requires some transformations of the variational problem and is too lengthy to be reproduced here.

References

[1] V. Bangert, Mather Sets for Twist Maps and Geodesic Tori, Preprint, Bonn, 1985, to appear in Dynamics Reported, vol. 1 (1988).

[2] V. Bangert, The existence of gaps in minimal foliations, to appear in Aequationes math. (1988).

[3] V. Bangert, A uniqueness theorem for $\mathbb{Z}^n$-periodic variational problems, to appear in Comm. Math. Helv. 1987.

[4] E. di Benedetto, N.S. Trudinger, Harnack Inequality for Quasi-Minima of Variational Integrals, Annales de l'Inst. H. Poincaré, Analyse Non-Linéaire, t. 1, 1984, 295-308.

[5] A. Celletti , L. Chierchia, On a rigorous stability result for bidimensional KAM tori, Forschungsinstitut für Mathematik, ETH Zürich, July 1987.

[6] J. Denzler, Mather sets for plane Hamiltonian systems, to appear ZAMP 1988.

[7] M. Giaquinta, E. Giusti, Quasi-minima, Ann. de l'Inst. Henri Poincaré, Analyse Non-Linéaire, t. 1, 1984, 79-107.

[8] M. Giaquinta, E. Giusti, Differentiability of Minima of Non-Differentiable Functionals, Inv. Math., t. 72, 1983, 285-298.

[9] M. Giaquinta, Multiple Integrals in the Calculus of Variations and Nonlinear Elliptic Systems, Ann. Math. Studies, t. 105, Princeton, N.J., 1983.

[10] G.A. Hedlund, Geodesics on a two-dimensional Riemannian manifold with periodic coefficients, Ann. Math., t 33, 1982, 719-739.

[11] M.R. Herman, Sur les courbes invariantes par les difféomorphismes de l'anneau, Astérisque, 103-104, 1983, 1-221.

[12] O.A. Ladyzhenskaya, N.N. Uraltseva, Linear and Quasilinear Elliptic Equations, Acad, Press, New York and London, 1968.

[13] J.N. Mather, Existence of quasi-periodic orbits for twist homeomorphisms of the annulus, Topology, t 21, 1982, 457-467.

[14] J.N. Mather, Destruction of invariant circles, Forschungsinstitut für Mathematik, ETH Zürich, June 1987.

[15] M. Morse, A fundamental class of geodesics on any closed surface of genus greater than one. Trans. Am. Math. Soc., t 26, 1924, 25-60.

[16] J. Moser, Recent Developments in the theory of Hamiltonian Systems, SIAM Review 28, Dec. 1986.

[17] J. Moser, Minimal solutions of variational problems on a torus, Ann. Inst. Henri Poincaré, Analyse Non-Linéaire, vol. 3, No. 3, 1986, 229-272.

[18] J. Moser, Breakdown of stability, Lecture Notes in Physics 247, Springer Verlag 1986, pp 492-518.

[19] D. Salamon and E. Zehnder, KAM-theory in configuration space, to appear in Ergodic Theory and Dynamical Systems 1988.

[20] J. Mather, Minimal Measures, Forschungsinstitut für Mathematik, ETH Zürich, 1987.

[21] E.A. Coddington and N. Levinson, Theory of Ordinary Differential Equations, Mc Graw Hill 1955, especially Chap. 17.

[22] J. Mather, Nonexistence of invariant Circles, Ergodic Theory and Dynamical Systems 4, 1984, 301-309.

VARIATIONAL METHODS IN NONLINEAR PROBLEMS

L. Nirenberg
Courant Institute
New York University
New York, NY 10012

I. In recent years variational methods have proved remarkably fruitful, and flexible, in attacking nonlinear problems. In a variational approach one tries to find solutions of a given equation by looking for stationary points of a (real) functional defined in the space in which the solution is to lie — the given equation is the Euler-Lagrange equation satisfied by a stationary point. The functional is often unbounded, from above and below, so one cannot look for maxima or minima. Instead one seeks saddle points by a min-max argument. In addition we may want to find multiple, perhaps infinitely many, stationary points.

This series of four lectures is the most elementary in this C.I.M.E. It is meant for people not in the field — to introduce to them a few of the arguments that currently are used for finding stationary points. We will start with some old results and lead up to some new ones — primarily concerning time-periodic solutions of systems of ordinary differential equations. I plan to include proofs or at least the main ideas of some of the proofs. Some computations will be left as exercises. An excellent introduction to the subject, and indeed for variational methods in general, is furnished by the 1974 C.I.M.E. lectures [12] of P. Rabinowitz. (In fact we begin with some material from there.) For those who wish to pursue the subject further we recommend also his more recent lectures [14]. We also mention [11].

We begin with a result which is familiar to most of the audience for finding a nontrivial stationary point of a given real C^1 function F defined in a Banach space X.

<u>Mountain Pass Lemma</u> (MPL). Assume there is an open neighborhood U of the origin in X, and a constant c_0 such that for some point $u_0 \in X \setminus U$,

$$F(u_0),\ F(0) < c_0 \leq F(u) \quad \forall u \in \partial U . \tag{1}$$

Then the following number $c \geq c_0$:

$$c = \inf_{p} \max_{u \in p} F(u) , \tag{2}$$

is a stationary value of F i.e. there is a stationary point of F where F equals c.

Here p represents any continuous path joining 0 to u_0 in X; we maximize F over p and then take infimum with respect to all possible paths. Since every such path p must cross ∂U, where we have $F \geq c_0$, we see that $\max_p F \geq c_0$. Think of F as representing height of land at a point u. Then 0 is a point in a valley U bounded by a mountain range ∂U. For any path p joining u_0 to 0, $\max_p F$ represents how high we have to go on that path. Taking infimum then minimizes this. Clearly c is the height of the lowest mountain pass crossing the mountain range. When we're at that mountain pass the earth is level, so F' vanishes.

This result is intuitively obvious however, as stated, it is false, even in finite dimensions. Here is a counterexample in $\mathbb{R}^2$. In the complex plane $\mathbb{C}$ consider the nonnegative function

$$F(z) = |e^z - 1|^2 .$$

Clearly F achieves the minimum, zero, at 0 and $2\pi i$.

Exercise: Prove that for $r > 0$ small

$$F(z) \geq c_0 > 0 \quad \text{for} \quad |z| = r .$$

On the other hand show that zero is the only critical value of F.

Thus F satisfies (1) but the conclusion of the MPL does not hold. One has to add an additional condition to MPL. A traditional one is the so-called Palais-Smale condition, a kind of compactness condition:

$(PS)_c$: Any sequence $\{u_i\}$ in X for which $F(u_i) \to c$ and $F'(u_i) \to 0$ strongly in X^*, has a strongly convergent subsequence $\{u_{i_j}\}$ in X.

Recall that the Frechet derivative $F'(u)$ of F at u represents a continuous linear functional on X, i.e. an element in the dual space X^*. We say that F satisfies (PS) if it satisfies $(PS)_c$ for all real c.

Correct MPL. The MPL formulated above holds under the additional condition $(PS)_c$ on F, where c is as in (2).

This follows immediately from the following form of (MPL):

Theorem 1. Let F satisfy the conditions of the (incorrect) MPL. Then there exists a sequence of points $\{u_i\}$ in X for which $F(u_i) \to c$ and $F'(u_i) \to 0$ strongly in X^*.

This is usually proved using a deformation argument due to M. Morse: Suppose the claim is false. Then there exist $\varepsilon, b > 0$ such that $||F'(u)|| > b$ in $\tilde{X} = \{u \in X \mid c - \varepsilon \leq F(u) \leq c+\varepsilon\}$. We may suppose ε

so small that 0 and u_0 are not in $\tilde{X}$. For simplicity we will suppose that X is a Hilbert space and that $F \in C^2$. Let $0 \le g(u) \le 1$ be a locally Lipschitz continuous function on X satisfying

$$g(u) = 1 \quad \text{if} \quad c - \frac{\varepsilon}{2} \le F(u) \le c + \frac{\varepsilon}{2} ,$$

$$g(u) = 0 \quad \text{outside } \tilde{X} .$$

Define the vector field (more or less the negative gradient of F)

$$V(u) = \begin{cases} -\, g(u) \dfrac{F'(u)}{||F'(u)||^2} & \text{in } \tilde{X} \\ \\ 0 & \text{outside } \tilde{X} \end{cases} \tag{3}$$

Clearly V is locally Lipschitz and $||V(u)|| \le b^{-1}$.

Consider the (essentially) negative gradient flow $v(t) = v(t,u)$ defined by

$$\frac{dv}{dt} = V(v) , \quad v\big|_{t=0} = u \quad \text{for} \quad u \in X .$$

There is a unique solution $v(t) = v(t,u)$ for $0 \le t < \infty$ which for each t yields a homeomorphism $u \mapsto v(t,u)$ of X onto X and this homeomorphism is identity outisde of $\tilde{X}$. Since

$$\frac{d}{dt} F(v(t)) = -\, g(v(t))$$

we see easily that the homeomorphism

$$\eta = v(\varepsilon)$$

maps the region $\{u \mid F(u) \le c + \varepsilon/2\}$ into the region $\{u \mid F(u) \le c-\varepsilon/2\}$. This is the Morse deformation of the identity map.

To complete the proof of Theorem 1 we note that by the definition of c there is a path p joining 0 to u_0 on which $F \le c + \varepsilon/2$. Consequently the deformed path $\eta(p)$ joins 0 to u_0 and on it $F \le c-\varepsilon/2$; contradiction.

In case X is a Banach space, and F is only in C^1, in (3) in $\tilde{X}$, we replace the gradient $F'(u)$ (which belongs to X^* rather than X) by a "pseudogradient" w which is a locally Lipschitz vector field in X satisfying

$$||w(u)|| \le 2||F'(u)||$$

$$\langle F'(u), w(u)\rangle \ge ||F'(u)||^2 ,$$

so that $||v(u)|| \ge ||F'(u)|| \ge b$.

In section IV we will use a second deformation result of Morse.

It's worth indicating another proof of Theorem 1 based on a simple and useful lemma due to I. Ekeland [6].

Ekeland's Lemma. Let M be a complete metric space, with distance between points (x,y) denoted by d(x,y). Let ψ be a real function on M with values in $(-\infty,\infty]$, $\psi \not\equiv \infty$, which is lower semicontinuous (l.s.c.), i.e.,

$$\{x \mid \psi(x) > c\} \text{ is open for every real } c.$$

Assume ψ is bounded from below. Then, given $\varepsilon > 0$, $\exists z \in M$ such that

$$\psi(z) < \inf_M \psi + \varepsilon$$

and

$$\psi(x) \geq \psi(z) - \varepsilon\, d(x,z) \quad \forall x \in M .$$

Ekeland's Lemma yields a slightly stronger form of Theorem 1 (where c is given by (2)):

Theorem 1'. Let F be a real C^1 function defined on a Banach space X. Suppose that for every continuous path p joining 0 and u_0 in X we have

$$F(0),\ F(u_0) < \max_p F .$$

Given $\varepsilon > 0$, there is a path $\bar{p}$ and a point on it, $\bar{u}$, where F achieves its maximum on $\bar{p}$, such that $F(\bar{u}) < c+\varepsilon$ and $\|F'(\bar{u})\| < \varepsilon$.

Idea of proof (Ekeland): Let C = set of continuous paths $p(t)$, $0 \leq t \leq 1$, $p(0) = 0$, $p(1) = u_0$, joining 0 to u_0. For every such path p define

$$\psi(p) = \max_t F(p(t)) .$$

With the uniform topology, C is a complete metric space and ψ is l.s.c. on C. Also $\psi(p) \geq F(p(0)) = F(0)$, and so is bounded from below. Clearly c of (2) equals $\inf_C \psi$.

Now take $\varepsilon > 0$ and apply Ekeland's Lemma; for simplicity we'll suppose again that X is a Hilbert space. There is a path $\bar{p}$ such that for any path p

$$\psi(p) \geq \psi(\bar{p}) - \frac{\varepsilon}{2} \max_t \|\bar{p}(t) - p(t)\| \tag{4}$$

and such that $\psi(\bar{p}) < c + \varepsilon$. Let

$$\bar{T} = \{t \in [0,1] \mid F(\bar{p}(t)) = \max_{\bar{p}} F\} .$$

So 0 and 1 are not in $\bar{T}$. The desired conclusion follows from the

Claim: $\exists\, \bar{t} \in \bar{T}$ such that $\|F'(\bar{p}(\bar{t}))\| < \varepsilon$.

Proof of Claim: Suppose false. Then $\forall\, \bar{t} \in \bar{T}$, $\|F'(\bar{p}(\bar{t}))\| \geq \varepsilon$. Define $\xi: [0,1] \to X$ as

$$\xi(t) = -\frac{F'(\bar{p}(t))}{\|F'(\bar{p}(t))\|} \quad \text{if} \quad t \in \bar{T} ,$$

$$\xi(0) = 0 , \quad \xi(1) = 1 ,$$

and with ξ extended continuously to $[0,1]$ to have norm ≤ 1.

Consider the comparison path

$$p_h(t) = \bar{p}(t) + h\,\xi(t) , \quad h > 0 \quad \text{small.}$$

Since $\|\xi(t)\| \leq 1$ we have by (4),

$$\psi(p_h) \geq \psi(\bar{p}) - \frac{\varepsilon}{2} h .$$

It follows that if $F(p_h(t))$ achieves its maximum at some point t_h then

$$F(p_h(t_h)) - F(\bar{p}(t)) \geq -\frac{\varepsilon}{2} h \quad \forall\, t .$$

If we let $h \to 0$ through a sequence, then, for some subsequence, $t_h \to \bar{t} \in \bar{T}$. So

$$F(\bar{p}(t_h) + h\,\xi(t_h)) - F(\bar{p}(t_h)) \geq -\frac{\varepsilon}{2} h ;$$

hence

$$-\frac{\varepsilon}{2} h \leq h \langle F'(\bar{p}(t_h)), \xi(t_h)\rangle + o(h) .$$

Dividing by h and letting $h \to 0$ we find

$$\langle F'(\bar{p}(\bar{t})), \xi(\bar{t})\rangle \geq -\varepsilon/2 ,$$

or

$$\|F'(\bar{p}(t))\| \leq \varepsilon/2 ;$$

contradiction. □

Many applications of MPL have been made: see the references cited above, as well as the paper [1] by A. Ambrosetti and Rabinowitz in which it first appeared in essentially this form.

We conclude this lecture with a simple application: we find at least two solutions for an elliptic boundary value problem (this was shown to us by K. C. Chang).

In Ω, a bounded domain in R^3 with smooth boundary $\partial\Omega$, consider the Dirichlet problem for a function u:

$$\begin{aligned} \Delta u + u^2 &= f(x) \geq 0 \quad \text{in } \Omega, \quad f \not\equiv 0 \text{ and smooth,} \\ u &= 0 \quad \text{on } \partial\Omega \end{aligned} \tag{5}$$

<u>Theorem 2.</u> <u>Problem</u> (5) <u>has at least two solutions</u>.

<u>Proof:</u> We obtain a first solution by constructing sub and super solutions of (5), i.e. functions $\underline{u} \leq \bar{u}$ vanishing on $\partial\Omega$ and satisfying in Ω,

$$\Delta \underline{u} + \bar{u}^2 - f \geq 0 ,$$

$$\Delta \bar{u} + \bar{u}^2 - f \leq 0 .$$

By a well known result in elliptic theory there then exists a solution u_0 of (5) satisfying

$$\underline{u} \leq u_0 \leq \bar{u} \quad \text{in} \quad \Omega .$$

We take $\bar{u} \equiv 0$, and $\underline{u}$ = the solution v of

$$\Delta v = f \quad \text{in} \quad \Omega , \qquad v = 0 \quad \text{on} \quad \partial\Omega .$$

Since $\Delta v = f \gneq 0$ we see by the maximum principle that

$$\underline{u} = v \leq \bar{u} = 0 .$$

To obtain a second solution we use the variational approach. Equation (5) is the Euler-Lagrange equation satisfied by any stationary point u of the functional

$$F(u) = \int_\Omega \frac{1}{2} |\nabla u|^2 - \frac{u^3}{3} + fu$$

for u in the Hilbert space $X = H_0^1$ the completion in the norm $\|u\| = [\int_\Omega |\nabla u|^2]^{1/2}$ of $C_0^\infty(\Omega)$.

<u>Exercise</u>: Using Sobolev's embedding theorem prove that in $X = H_0^1$ the function F is of class C^1 and satisfies (PS).

<u>Claim</u>: F has a strict local minimum at u_0. In fact for $\|w\| = r > 0$ small,

$$F(u_0 + w) - F(u_0) \geq \frac{1}{4} r^2 . \tag{6}$$

But for $v \in H^1$ satisfying $\int v^3 > 0$ we have $F(tv) < 0$ for t large. Using (6), it then follows from (correct) MPL that F has another stationary point in addition to u_0.

The claim is easily proved:

$$F(u_0+w) - F(u_0) = \int \nabla u \cdot \nabla w + \frac{1}{2} |\nabla w|^2 - u_0^2 w - u_0 w^2 - \frac{w^3}{3} + fw$$

$$= \int \frac{1}{2} |\nabla w|^2 - u_0^2 w - u_0 w^2 ,$$

(since u_0 is a solution of (5), and so a stationary point of F)

$$\geq \frac{1}{2} \|w\|^2 - \int \frac{w^3}{3} .$$

By Sobolev's embedding theorem we have the control

$$\int w^3 \leq C \|w\|^3.$$

So

$$F(u_0+w) - F(u_0) \geq \frac{1}{4} \|w\|^2 = \frac{1}{4} r^2$$

if $\|w\| = r$ small.

We have obtained two solutions. Using standard elliptic regularity theory one easily establishes that these are smooth.

II. In recent years, following the classical work of L. Lyusternik and L. Schnirelman [8], people have developed techniques for finding multiple (possibly infinitely many) stationary points for functionals in case the functional is invariant under some group action; see [3], [14]. A classic example is the following. Consider a C^1 real function F defined on the unit sphere S^{n-1} in $\mathbb{R}^n$. It has at least two stationary points, max and min, but need not have any more. On the other hand if F is even, $F(-x) = F(x)$, it has at least n pairs of stationary points (see [8], or [14] Theorem 7.1). The fact that F is invariant under the $\mathbb{Z}_2$ action generated by $x \longmapsto -x$ on S^{n-1} yields this multiplicity. One way of proving this is to observe that F is well defined on projective space which has a lot of nontrivial topology.

Another proof may be based on a topological result:

<u>Borsuk-Ulam Theorem:</u> <u>Let</u> U be a bounded symmetric open neighbourhood of the origin in $\mathbb{R}^n$. Suppose f is an odd map of $\partial\Omega$ <u>into</u> $\mathbb{R}^d \setminus \{0\}$. <u>Then</u> $d \geq n$.

Much of the recent work on multiple stationary points for variational problems treats functionals which are invariant under certain group actions such as S^1 action or $\mathbb{Z}_p$ action. Min-max arguments are used to obtain stationary values:

$$c = \sup_{A \in \Sigma} \inf_{A} F(u) . \tag{7}$$

Here Σ represents some class of closed sets A (in the space in which we are working) each of which is invariant under the group action. For each problem one has to find a suitable class Σ of sets with which to work. One also introduces a notion of "size" of these sets, or "index". Rabinowitz and V. Benci have played central roles in these developments (see [14]).

In the remainder of these lectures we will illustrate some of these ideas by finding multiple time-periodic solutions of systems of ordinary differential equations. In this lecture we will deal with second order equations and, subsequently, with more general Hamiltonian systems for which we seek, so-called subharmonic solutions (to be expanded). There we will use $\mathbb{Z}_p$ action.

Consider a system for vector valued (in $\mathbb{R}^n$) functions $x(t)$:

$$\ddot{x} + V_x(t,x) = 0 \tag{8}$$

with V a real smooth function defined in $\mathbb{R}\times\mathbb{R}^n$ and periodic in t of period τ, which we take to be 2π, and satisfying

$$0 < V(t,x) \leq \theta x\cdot V_x(t,x) \ , \quad 0 < \theta < \frac{1}{2} \ , \quad \text{for } |x| \text{ large and all } t. \tag{9}$$

The condition implies that for some $C, C' > 0$,

$$V(t,x) \geq C|x|^{1/\theta} - C' \ . \tag{9'}$$

We may assume

$$V(t,x) \geq 0 \ . \tag{10}$$

In this lecture we will prove the following

Theorem 3. Assume that $V = V(x)$ is independent of t and satisfies (9). Then for every $\tau > 0$, equation (8) has infinitely many different solutions with period τ.

Since V is independent of t the problem is translation invariant (under translation of t). Two solutions x,y are called different if $x(t + \text{const}) \neq y(t)$ $\forall$ constant. In looking for solutions of period τ we need only consider translations (mod τ) i.e. the S^1-action $(T_s x)(t) = x(t+s)$, $0 \leq s \leq \tau$.

Note that if V has quadratic growth the result may not hold. For example if $V = |x|^2$, the period of any nontrivial solution must be an integral multiple of 2π.

Theorem 3 was proved independently by Benci [3] and by Rabinowitz [13] using the S^1-invariance and index theory. Subsequently, A. Bahri and H. Berestycki [2] treated (in fact more general cases)

$$\ddot{x} + V_x(x) = f(t) \quad \text{having period } \tau. \tag{8'}$$

They proved

Theorem 3'. With V as in Theorem 3, and with $f \in L^2(\mathbb{R},\mathbb{R}^n)$ of period τ, there exist infinitely many solutions of (8)' with period τ.

We are going to give a rather simple proof of Theorem 3. It is a modification by G. Tarantello of an argument in Benci, D. Fortunato [4]. The proof by Theorem 3' in [2] is much more elaborate. It would be very worthwhile to find a simpler proof of it. In addition we would like to call attention to the following questions.

Problem 1. Does Theorem 3 hold if (9) is replaced by (9)' ?

Indeed the additional condition (9) is used only in verifying (PS) for the corresponding functional.

Problem 2. Under condition (9), does (8) have infinitely many solutions of period τ ?

Our proof of Theorem 3 will make use of the following

Generalized MPL. *Let F be a real C^1 function defined in a Banach space X and satisfying* (PS). *Let* U *be a bounded open set lying in a finite dimensional subspace* E *of* X *with*

$$F\big|_{\partial U} \le c_0 \; . \tag{11}$$

Assume that there is a constant $c_1 > c_0$ *such that for every continuous map*

$$h: \bar{U} \to X \quad \text{with} \quad h\big|_{\partial U} = \mathrm{Id} \; , \tag{12}$$

we have

$$\max_{\text{Image of } h} F \ge c_1 \; . \tag{13}$$

Then the following number

$$c := \inf_h \max_{x \in \bar{U}} F(h(x)) \ge c_1 \tag{14}$$

is a stationary value of F.

Here the infimum is taken over all continuous maps as in (12).

The proofs of MPL in the previous section extend to this generalized form.

We will first discuss the general equation (8) with $V(t,x)$ and then prove Theorem 3 for V independent of t. Associated with (8) is the variational expression for vector valued functions $x(t)$ of period τ (taken to be 2π):

$$F[x] = \int_0^{2\pi} [\, \tfrac{1}{2} |\dot{x}|^2 - V(t, x(t))]\, dt \; . \tag{15}$$

A stationary point x of (15) is a 2π-periodic solution of (8).

We'll have need of the following

Calculus Lemma: If $x(t)$ has period 2π, $\int_0^{2\pi} x(t)\,dt = 0$ and $|x|_{L^\infty} = 1$, then

$$\frac{1}{4\pi} \int_0^{2\pi} |\dot{x}|^2\, dt \ge \frac{1}{4} \left(\sum_1^\infty j^{-2} \right)^{-1} =: a \; .$$

This is easily proved using Fourier series expansion:

$$x(t) = \sum_{j \ne 0} a_j e^{ijt} \; , \quad a_j \in \mathbb{C}^n \; .$$

For

$$1 = |x|_{L^\infty} \le \sum |a_j| \le \sqrt{\sum_{j \ne 0} j^2 |a_j|^2}\; \sqrt{2 \sum_1^\infty j^{-2}} \; ,$$

while by Parseval's identity,

$$\int |\dot{x}|^2\, dt = 2\pi \sum_{j\neq 0} j^2 |a_j|^2 .$$

The result follows.

Here is a result for (8) $(\tau = 2\pi)$ if V is suitably small.

Theorem 4. Suppose (9) and (10) hold, and that

$$K := \max_{\substack{t \\ |x|\leq 1}} V(t,x) < a . \tag{16}$$

Then (8) has a nonconstant periodic solution with period 2π.

A simple trick yields the

Proof of Theorem 3: We construct nonconstant solutions with period $2\pi/k$ for large integral k. Set

$$t = \frac{s}{k}$$

Then in terms of the s variable we look for solutions x in s of the corresponding equation

$$x_{ss} + \frac{1}{k^2} V_x(x) = 0 , \qquad x \text{ of period } 2\pi \text{ in } s .$$

This is of the form (8) with a new $\tilde{V} = \frac{1}{k^2} V$. For k sufficiently large this V satisfies the conditions of Theorem 4 and hence has a nonconstant solution x(s) of period 2π. In terms of the $t = s/k$ variable its period is $2\pi/k$. It may have smaller minimal period $2\pi/k'$ with $k' \geq k$. Choose an integer $k_2 > k'$ and repeat the argument: we obtain a solution of period $2\pi/k_2$ so it must be different from the first one. Continuing in this way we obtain infinitely many different solutions. □

Proof of Theorem 4: We will obtain a nonconstant stationary point of the functional F in (15) working in the Banach space $X = H^1$ of 2π-periodic (vector valued) functions.

Exercise: Show that F is of class C^1 in X and satisfies (PS).

As we mentioned, the strong form (9) of (9)' enters only in verifying (PS).

In X we will apply the generalized MPL above with E the (n+1)-dimensional subspace of X:

$$E = \left\{ x(t) = x_0 + \beta e_1 \cos t \;\middle|\; x_0 \in \mathbb{R}^n,\ \beta \in R,\ e_1 = (1,\dots,0) \right\}$$

and with U the open cylinder

$$U = \left\{ x(t) = x_0 + \beta e_1 \cos t \;\middle|\; |x_0| < R,\ 0 < \beta < R,\ R \text{ large} \right\} .$$

For x in E we have

$$F[x] = \frac{1}{2} \beta^2 \int_0^{2\pi} \sin^2 t - \int_0^{2\pi} V(t, x(t)) = \frac{\pi}{2} \beta^2 - \int V(t,x) .$$

Claim: For R large we have $F \le 0$ on ∂U.

Proof: On the base of the cylinder, $\beta = 0$, Since $V \ge 0$ we have $F(x) \le 0$ there.

For $x = x_0 + \beta e_1 \cos t$ in E we have

$$\int_0^{2\pi} |x|^2 = 2\pi |x_0|^2 + \pi\beta^2 .$$

In addition, by (9)',

$$\begin{aligned}\int_0^{2\pi} V(t, x(t)) &\ge C \int |x(t)|^{1/\theta} - C' \\ &\ge C_1 \left(\int |x|^2\right)^{1/2\theta} - C'\end{aligned}$$

by Hölder's inequality — with a suitable constant C_1. Hence

$$F[x] \le \frac{\pi}{2}\beta^2 - C_1 (2\pi |x_0|^2 + \pi\beta^2)^{1/2\theta} + C' . \tag{17}$$

Thus on the lateral surface of the cylinder, i.e., where $|x_0| = R$ and $0 \le \beta \le R$, we have

$$\begin{aligned}F[x] &\le \frac{\pi}{2} R^2 - C_1 (2\pi\ R^2)^{1/(2\theta)} + C' \\ &\le 0 \qquad\qquad \text{for } R \text{ large}\end{aligned}$$

since $1 > 2\theta$.

Similarly, on the remaining boundary of U, where $\beta = R$ we have from (17)

$$\begin{aligned}F[x] &\le \frac{\pi}{2} R^2 - C_1\ 2\pi (\pi R^2)^{1/(2\theta)} + C' \\ &\le 0 \qquad\qquad \text{for } R \text{ large} .\end{aligned}$$

The claim is proved and thus (11) holds with $c_0 = 0$.

To apply the generalized MPL we will verify condition (13) for every permissible map, with $c_1 > 0$. Then $c \ge c_1 > 0$ in (14) is a critical value, and a corresponding critical point must be nonconstant since for any constant critical point x we have $F(x) = -\int V(t,x)dt \le 0$.

To establish (13) it suffices to prove the

Claim: The image of any admissible map h, i.e. satisfying (12), contains a point y satisfying

$$\int_0^{2\pi} y(t)\ dt = 0 \quad \text{and} \quad |y|_{L^\infty} = 1 .$$

For then we have

$$F[y] = \int \left[\frac{1}{2} |\dot{y}|^2 - V(t,y(t))\right] dt \ \ge\ 2\pi(a-K) = c_1 > 0,$$

by the Calculus Lemma and (16).

Proof of Claim: We wish to find $x = x_0 + \beta e_1 \cos t$ in U so that

$$\int h(x)(t)\,dt = 0 \quad \text{and} \quad |h(x)|_{L^\infty} - 1 = 0 .$$

These are (n+1) equations for (n+1) unknowns and we will show they are solvable with the aid of Brouwer degree theory.

The map of $\bar{U}$ into R^{n+1} given by

$$(x_0,\beta) \overset{f}{\longmapsto} \left(\int h(x)(t)\,dt\ ,\ |h(x)|_{L^\infty} - 1\right)$$

does not vanish on ∂U, for on ∂U we have h = Id so the map is

$$(x_0,\beta) \longmapsto (2\pi x_0,\ |x_0 + \beta\, e_1 \cos t|_{L^\infty} - 1) .$$

At a point where this vanishes we have $x_0 = 0$ and $\beta = 1$, and this is not on ∂U. So the degree of the map f at the origin deg(f,U,0) is well defined and is the same as that of the map (which agrees with f on ∂U)

$$(x_0,\beta) \overset{g}{\longmapsto} (2\pi x_0,\ |x_0+\beta\, e_1 \cos t|_{L^\infty} - 1) .$$

To complete the proof we show that the degree of this map is one. Namely deform the map g via a one-parameter family of maps g_λ , $0 \le \lambda \le 1$,

$$g_\lambda(x_0,\beta) = (2\pi x_0, |\lambda\, x_0+\beta\, e_1 \cos t|_{L^\infty} - 1) ,$$

to the map $g_0 = (2\pi x_0,\beta-1)$. Note that if $g_\lambda(x_0,\beta) = 0$ then $x_0 = 0$ and $\beta = 1$ so $(x_0,\beta) \notin \partial U$. Thus the degree

$$\deg(f,U,0) = \deg(g_0,U,0).$$

But g_0 is essentially the identity map and its degree is one. □

III. The last two lectures are concerned with a result on time-periodic solutions for a class of Hamiltonian systems. Here z(t) is a vector valued function of t with values in R^{2N} and the system takes the form

$$J\dot{z} = H_z(z,t) \tag{18}$$

where H is a real continuous function in $\mathbb{R}^{2N} \times \mathbb{R}$ which is periodic in t of period τ, which we take to be 2π; J is the 2N by 2N matrix

$$J = \begin{pmatrix} 0 & Id \\ -Id & 0 \end{pmatrix} .$$

If H is independent of t the system is called autonomous. We are going to treat the nonautonomous case. In fact we will assume that 2π is the smallest t-period of H in the following precise sense:

(H1): If for some $z(t)$ with minimal period $2\pi q$, $q > 0$ rational, the (vector) function $H_z(t,z(t))$ has period $2\pi q$ then q = integer.

This clearly holds in case, say,

$$H(z,t) = \hat{H}(z)\ a(t)\ ,$$

where $a(t) > 0$ has minimal period 2π.

We will present a result of R. Michalek and G. Tarantello [10] concerning the existence of solutions with smallest period $2\pi p$ where $p > 1$ is a given integer. Such solutions are called subharmonic.

By now many people have contributed to the problem of finding one or more solutions with given minimal period, but mainly in the autonomous case. In that case one obtains multiple periodic solutions by using the fact that the problem is invariant under time shift: $t \mapsto t+c$; because of periodicity this is an S^1-action. Fadell and Rabinowitz, and Benci, have extended the ideas of [8] and introduced suitable "index theories" (see [3] and references in [14]). These ideas have been extended to $\mathbb{Z}_p$-action, which we will use here, independently by several people; by R. Michalek [9] and Wang, Zhi-Quang [16-17] (see other references here). Ekeland suggested to Michalek and Tarantello to try to use $\mathbb{Z}_p$-action in order to obtain subharmonic solutions. We now describe one of their results. Assume

(H2): H is smooth for $z \neq 0$, convex in z, and has uniform subquadratic growth

$$\frac{a_1}{\beta}\,|z|^\beta \le H(z,t) \le \frac{a_2}{\beta}\,|z|^\beta \qquad 1 < \beta < 2 \qquad a_1 > 0 \tag{19}$$

Theorem 5 ([10]). Let s_p be the smallest prime factor of p and assume that for some integer $k \geq 1$,

$$\frac{a_2}{a_1} < \left(\frac{2s_p}{k(k+1)}\right)^{\beta/2} . \tag{20}$$

Assume also (H1) and (H2), then (18) has at least kN different solutions with minimal period $2\pi p$.

The result means that if s_p is large then there are many solutions with minimal period $2\pi p$. Two solutions z,w are called different if $z(t+2\pi j) \not\equiv w(t)$ $\forall$ integer j.

Theorem 5 is rather special — the condition (H2) is very strong. In fact recently Tarantello [15] has obtained more general results under much weaker conditions than (H2), also in case H is superquadratic in z. But the proofs are considerably more complicated than that of Theorem 5. Since the latter also illustrates a number of current ideas in the subject we have decided to present it here. For simplicity we'll suppose H is strictly convex in z.

The standard variational expression associated with $2\pi p$-periodic solutions of (18) is

$$F(z) = \int_0^{2\pi p} [\tfrac{1}{2}(J\dot{z},z) - H(z(t),t)]\, dt ;$$

(18) is the Euler equation for any stationary point z of period $2\pi p$. This functional is of course unbounded from above and below.

We now start the proof of Theorem 5.

<u>Step 1</u>. The first step in the proof is to replace F by a different variational problem, the

<u>Dual Variational Problem</u> (F. H. Clarke - Ekeland [5]):

We first rewrite the equation (18) in a different form by essentially inverting the linear operator $J\frac{d}{dt}$ occurring in (18), and the nonlinear one $z \mapsto H_z(z,t)$, with t thought of as a parameter.

Consider the gradient map

$$z \longmapsto u = H_z(z,t) . \tag{21}$$

For H strictly convex in z this map is one to one and its inverse is given by

$$z = H_u^*(u,t) , \tag{21'}$$

where H^* is the dual convex function to H:

$$H^*(u,t) = \sup_z ((u,z) - H(z,t)) .$$

From (19) one finds that H^* satisfies

$$b_1|u|^\alpha \le H^*(u,t) \le b_2|u|^\alpha , \qquad \frac{1}{\alpha} + \frac{1}{\beta} = 1 \tag{19'}$$

with

$$b_1 = \frac{1}{\alpha} a_2^{1-\alpha} , \qquad b_2 = \frac{1}{\alpha} a_1^{1-\alpha} .$$

Thus H^* grows superquadratically.

Equation (18) now takes the form

$$J\frac{d}{dt} z - u = 0 .$$

Let $K = (J\frac{d}{dt})^{-1}$; this is well defined on functions u(t) of period $2\pi p$ with

$$\int_0^{2\pi p} u(t)\, dt = 0 \tag{20}$$

and

$$\int_0^{2\pi\mu} (Ku)(z)\, dt = 0 .$$

Applying K we find, since $\int z\, dt$ need not be zero,

$$z - Ku = \text{const. vector}$$

or, from (21)',

$$Ku - H_u^*(u,t) = \text{const.} \tag{18)'$$

This is the new form of equation (18). If we have a $2\pi p$-periodic solution u of it satisfying (20) then we may reverse these steps and obtain a solution z of (18).

The natural variational expression that goes with (18)' for u satisfying (20) is

$$I[u] = \int_0^{2\pi p} [\tfrac{1}{2}(Ku,u) - H^*(u(t),t)]\, dt . \tag{21}$$

This is the dual variational expression of Clarke and Ekeland [5] with which we will work. (This procedure has proved useful for other problems including elliptic and some hyperbolic equations.)

In view of (19)' we see that a natural space in which to work is

$$X = L_0^\alpha = \{u \text{ of period } 2\pi p \text{ in } L^\alpha \mid \int_0^{2\pi p} u = 0\} .$$

A word about the operator K. It is a simple integral (compact) operator with spectrum

$$\sigma(K) = \frac{p}{j}, \quad j = \text{integer} \neq 0 . \tag{22}$$

E_j, the eigenspace of K corresponding to the eigenvalue p/j, is 2N-dimensional, and, in complex form, with

$$u_j + i\, u_{N+j} = v_j , \quad j = 1,\dots,N,$$

has as general element $v = (v_1,\dots,v_N)$:

$$v = \zeta\, e^{-ijt/p} , \quad \zeta \in \mathbb{C}^N . \tag{23}$$

<u>Exercise</u>: Prove that in $X = L_0^\alpha$, I is of class C^1 and satisfies (PS).

With the aid of (19)' one sees that I is bounded from above. It follows from the following exercise that there is a maximum point of I.

<u>Exercise</u>: Let I be a C^1 real function on a Banach space which satisfies (PS) and is bounded from above. Prove that I achieves a maximum.

<u>Step 2</u>. We now know that I has a stationary point, and so that (18) has a solution of period $2\pi p$. However we are looking for many solutions.

In addition they are to have minimal period $2\pi p$. How does one ensure that? [10] follows an idea of M. Girardi and M. Matzeu [7]. It makes use of the following

Calculus Lemma. Suppose $u \in L_0^\alpha$ has period $2\pi p/m$, with m an integer > 1. Then

$$I[u] \leq \frac{A(p)}{m^{\alpha/(\alpha-2)}} ,$$

with

$$A(p) = (\frac{1}{2} - \frac{1}{\alpha}) 2\pi p \cdot p^{\alpha/(\alpha-2)} \; a_2^{2(\alpha-1)/(\alpha-2)} .$$

The proof is left as an exercise: One expands u in a Fourier series, essentially, the eigenvectors of K, and then uses (22) and (19)' in a straightforward way.

The lemma is employed in the following way. Suppose u is a solution of (18)' with period $2\pi p/m$, m = integer > 1. Then we have a corresponding solution z of (18) with the same period. From (18) it follows with the aid of (H1) that p/m is an integer. Hence $m \geq s_p$ = the smallest prime factor of p. Thus from the Calculus Lemma we infer that

$$I[u] \leq \frac{A(p)}{s_p^{\alpha/(\alpha-2)}} =: B . \tag{24}$$

It follows that if u is a stationary point of I with I(u) > B then u has minimal period $2\pi p$. So we will seek such u.

IV. Step 3. We wish to use a min-max procedure to obtain many critical points u with I(u) > B. It will be based on the fact that the problem is invariant under the simple $\mathbb{Z}_p$-action: T_j , j = 1,...,p with T_p = Id and $T_j = T^j$, where

$$(Tu)(t) = u(t + 2\pi) ,$$

i.e., the equations (18), (18)' are invariant, as are the functionals F and I. Now we introduce some topology.

Associated with the action is a $\mathbb{Z}_p$-index theory (related to that of Benci for S^1-action in [3]) defined for closed sets A in L_0^α which are invariant under the action of T. Such sets are called admissible.

$\mathbb{Z}_p$-index theory: For every admissible set A we define an index i = integer ≥ 0, or $+\infty$, as follows.

Definition.

$$i(A) = 0 \quad \text{if} \quad A = \emptyset .$$

For $A \neq \emptyset$, i(A) = the smallest positive integer n such that there exist a continuous map

$$h = (h_1, \ldots, h_n) : A \to \mathbb{C}^n \setminus \{0\} ,$$

and nonzero integers $k_1, \ldots, k_n$, relatively prime to p, such that h is equivariant in the sense

$$h_j(Tu) = e^{ik_j 2\pi/p} h_j(u) , \quad j = 1, \ldots, n, \quad u \in A .$$

If no such map and integers exist we define

$$i(A) = +\infty .$$

A natural question is: Are there admissible sets with large index? Here is an example (which we will use).

<u>Theorem 6</u> ([9]). <u>In</u> $E_1 \oplus \ldots \oplus E_\ell$ <u>with</u> $\ell < s_p$ <u>the ellipsoid</u>

$$\Gamma_\ell : \begin{cases} v(t) = \sum_{j=1}^{\ell} \zeta_j e^{-ijt/p} , & \zeta_j \in \mathbb{C}^N \\ \text{with} & \\ \sum_1^\ell a_j |\zeta_j|^2 = a = \underline{\text{constant}} > 0 , & a_1, \ldots, a_\ell > 0 , \end{cases} \tag{25}$$

<u>has index</u> ℓN.

This is proved by an extension of the Borsuk-Ulam theorem mentioned earlier for this $\mathbb{Z}_p$-action.

We now describe the min-max argument. Let Σ denote the collection of admissible sets A in X. Set

$$c_\ell = \sup_{\substack{A \in \Sigma \\ i(A) \geq \ell}} \inf_{u \in A} I(u) , \quad \ell = 1, 2, \ldots .$$

From the definition we see that

$$c_1 \geq c_2 \geq \ldots .$$

<u>Proposition</u>. <u>Each</u> c_ℓ <u>is a critical value of</u> I.

This is proved as in the first section, with the aid of the Morse deformation argument, following (essentially) the gradient flow of I. I'm cheating here since X is not a Hilbert space; one should use pseudogradient — but never mind. Since I is invariant under the action T, one can ensure that the deformation of any admissible set A remains admissible. Also its index does not change. So the earlier proof can be readily made to work here.

<u>Step 4</u>. Recall that we want not only many critical points, but also having critical values > B (see (24)). To this end we have the following

<u>Calculus Lemma</u>. With k as in condition (20), on the ellipse Γ_k given by (25) with $a_j = 1/j$ and

$$a = p^{\frac{2}{\alpha-2}} a_1^{2(\alpha-1)/(\alpha-2)} \left(\frac{2}{k(k+1)}\right)^{\alpha/(\alpha-2)} ,$$

the inequality

$$I(u) > B$$

holds.

The proof is a fairly straightforward computation and uses (20), (19)', (22), (24).

It follows from this lemma and Theorem 6 that

$$c_1 \geq \dots \geq c_{kN} > B . \tag{26}$$

Thus we have obtained kN critical values with the property that they are all greater than B. So if they are all different we have indeed kN different solutions of (18) with minimal period $2\pi p$. However some of the c_j for $j \leq kN$ may be the same. Further work is required. It is analogous to arguments in [12], [14] and [3], but for those who are just now being exposed to this subject I'll sketch the main points.

We will make use of some properties of the index. These are easily proved (in just the same way as for Benci's S^1-index [3] and for genus [12]). The sets considered are assumed to be admissible.

Some propertes of the index. (i) $i(\overline{A \setminus B}) \geq i(A) - i(B)$.

(ii) Let K be compact and admissible. For $\delta > 0$ sufficiently small, the closed δ-neighbourhood of K, $N_\delta(K)$ is admissible and

$$i(N_\delta(K)) = i(K) .$$

In addition we will use a second deformation argument of Morse in an equivariant form relative to the action T. It is proved just as in [14] for S^1-action. For c real, set

$$K_c = \{\text{critical points } u \text{ of } I \text{ with } I(u) = c\}$$

$$A_c^+ = \{u \mid I(u) \geq c\}$$

$$A_c^- = \{u \mid I(u) \leq c\} .$$

By (PS), K_c is compact.

$\mathbb{Z}_p$-Deformation Theorem. For any neighbourhood U of K_c there exists $\bar{\varepsilon} > 0$ such that for any $0 < \varepsilon < \bar{\varepsilon}$ there is a T-equivariant homeomorphism η of X satisfying

$$\eta(A_{c-\varepsilon}^+ \setminus U) \subset A_{c+\varepsilon}^+ .$$

Suppose now that for some $\ell \geq 1$, $r > 0$, $\ell + r \leq kN$,

$$c_\ell = c_{\ell+1} = \dots = c_{\ell+r} = c ,$$

so $c > B$. We will show that in this case, K_c, the set of critical points with critical value c, contains infinitely many points. The proof is based on two lemmas.

Lemma A: Under the preceding conditions, $i(K_c) > r$.

Lemma B: If $c > B$ and $i(K_c) > 1$ then K_c contains infinitely many points.

These yield the desired result that K_c has infinitely many points, and completes the proof of Theorem 5.

Proof of Lemma A: Suppose $i(K_c) \leq r$. Using property (ii) of the index, take $\delta > 0$ so small that

$$i(N_\delta(K_c)) = i(K_c)$$

By the $\mathbb{Z}_p$-deformation theorem, for $\varepsilon > 0$ small, we can find a T-equivariant homeomorphism $\eta: X \to X$ satisfying

$$\eta(A^+_{c-\varepsilon} \setminus N_\delta(K_c)) \subset A^+_{c+\varepsilon} .$$

But since $c = c_{\ell+r}$ there exists A^ε with $i(A^\varepsilon) \geq \ell+r$ and

$$I \geq c-\varepsilon \quad \text{on} \quad A^\varepsilon .$$

But then

$$\text{for } u \in \eta(\overline{A^\varepsilon \setminus N_\delta(K_c)}) \quad \text{we have} \quad I(u) \geq c+\varepsilon , \tag{27}$$

and

$$i(\eta(\overline{A^\varepsilon \setminus N_\delta(K_c)})) = i(\overline{A^\varepsilon \setminus N_\delta(K_c)}) ,$$

since η is a T-equivariant homeomorphism,

$$\geq i(A^\varepsilon) - i(N_\delta(K_c)) \geq \ell .$$

This and (27) contradict the fact that $c = c_\ell$. □

Proof of Lemma B: Suppose K_c contains only finitely many points $u_1, \ldots, u_s$ and their translates $T^j u_m$, $0 < j < p$, $i \leq m \leq s$. Since $c > B$ they all have minimal period $2\pi p$; and so are distinct.

The map $h: K_c \to \mathbb{C} \setminus \{0\}$ given by

$$h(T^j u_m) = e^{ij2\pi/p} , \quad j = 0, \ldots, p-1$$

is well defined and satisfies

$$h(T(T^j u_m)) = h(T^{j+1} u_m) = e^{i2\pi/p} h(T^j u_m) .$$

Consequently $i(K_c) = 1$; contradiction. □

Acknowledgement: The work was partly supported by ONR N00014-85-K-0195 and ARO, DAA-29-84-K-0150.

References

[1] A. Ambrosetti, P. H. Rabinowitz. Dual variational methods in critical point theory and applications, J. Func'l. Anal. 14 (1973) p. 349-381.

[2] A. Bahri, H. Berestycki. Existence of forced oscillations for some nonlinear differential equations, Comm. Pure Appl. Math. 37 (1984) p. 403-442.

[3] V. Benci. A geometrical index for the group S^1 and some applications to the study of periodic solutions of ordinary differential equations, Comm. Pure Appl. Math. 34 (1981) p. 393-432.

[4] V. Benci, D. Fortunato. Un teorema di molteplicità per un' equazione ellittica nonlineare su varietà simmetriche, Metodi asintotici e topologici in problemi differenziali non lineari, Ist. Mat. Univ. dell'Aquila 1981. (Ed. L. Boccardo and A. M. Micheletti), Pitagora Ed. Bologna.

[5] F. H. Clarke, I. Ekeland. Hamiltonian trajectories having prescribed minimal period, Comm. Pure Appl. Math. 33 (1980) p. 103-116.

[6] I. Ekeland. On the variational principle, J. Math. Anal. Appl. 47 (1974) p. 324-353.

[7] M. Girardi, M. Matzeu. Solutions of minimal period for a class of nonconvex Hamiltonian systems and application to the fixed energy problem, Nonlin. Anal. TMA 10 (1986) p. 371-382.

[8] L. Lyusternik, L. Schnirelman. Méthodes topologiques dans les problèmes variationnels, Hermann et Cie, Paris, 1934.

[9] R. Michalek, Multiplicity results for differential equations with symmetry, Ph.D. Thesis, New York University, 1986.

[10] R. Michalek, G. Tarantello. Subharmonic solutions with prescribed minimal period for nonautonomous Hamiltonian systems, J. Diff. Eq's., to appear.

[11] L. Nirenberg. Variational and topological methods in nonlinear problems, Bull. Amer. Math. Soc. 4 (1981) p. 267-302.

[12] P. H. Rabinowitz. Variational methods for nonlinear eigenvalue problems, Eigenvalues of Nonlinear Problems (G. Prodi, Ed.) C.I.M.E. Ed. Cremonese, Roma, 1975, p. 141-195.

[13] P. H. Rabinowitz. Periodic solutions of Hamiltonian systems, Comm. Pure Appl. Math. 31 (1978) p. 157-184.

[14] P. H. Rabinowitz. Minimax methods in critical point theory with applications to differential equations, Conf. Board of Math. Sci. Reg. Conf. Ser. in Math., No. 65, Amer. Math. Soc., 1986.

[15] G. Tarantello. Subharmonic solutions for Hamiltonian systems via a $\mathbb{Z}_p$-index theory, Submitted to Annali della Scuola Norm. Sup. di Pisa.

[16] Z. Q. Wang. A $\mathbb{Z}_p$-index theory; to appear.

[17] Z. Q. Wang. A $\mathbb{Z}_p$-Borsuk-Ulam theorem; to appear.

Variational Theory for the Total Scalar Curvature Functional for Riemannian Metrics and Related Topics

Richard M. Schoen
Mathematics Department
Stanford University
Stanford, CA 94305

The contents of this paper correspond roughly to that of the author's lecture series given at Montecatini in July 1987. This paper is divided into five sections. In the first we present the Einstein–Hilbert variational problem on the space of Riemannian metrics on a compact closed manifold M. We compute the first and second variation and observe the distinction which arises between conformal directions and their orthogonal complements. We discuss variational characterizations of constant curvature metrics, and give a proof of Obata's uniqueness theorem. Much of the material in this section can be found in papers of Berger–Ebin [3], Fischer–Marsden [8], N. Koiso [14], and also in the recent book by A. Besse [4] where the reader will find additional references.

In §2 we give a general discussion of the Yamabe problem and its resolution. We also give a detailed analysis of the solutions of the Yamabe equation for the product conformal structure on $S^1(T) \times S^{n-1}(1)$, a circle of radius T crossed with a sphere of radius one. These exhibit interesting variational features such as symmetry breaking and the existence of solutions with high Morse index. Since the time of the summer school in Montecatini, the beautiful survey paper of J. Lee and T. Parker [15] has appeared. This gives a detailed discussion of the Yamabe problem along with a new argument unifying the work of T. Aubin [1] with that of the author.

§3 contains an a priori estimate on arbitrary (nonminimizing) solutions of the Yamabe problem in terms of a bound on the energy. The estimate applies uniformly to solutions of the subcritical equation, and implies that solutions of the subcritical equation converge in C^2 norm to solutions of the Yamabe equation. These estimates hold on manifolds which are not conformally diffeomorphic to the standard sphere. We present here the result for locally conformally flat metrics. This estimate has not appeared in print prior to this paper although we discovered it some time ago.

In §4 we discuss asymptotically flat manifolds and total energy for n-dimensional manifolds. We discuss the positive energy theorems which are needed for the Yamabe problem. We give a detailed n-dimensional proof of the author's work with S.T. Yau [25], [26] which proves the positive energy theorem through the use of volume minimizing hypersurfaces. The proof we give works for $n \leq 7$ in which dimensions we have complete regularity of volume minimizing hypersurfaces. Along with the locally conformally flat case which is treated in [29], this covers all cases which are used in the resolution of the Yamabe problem. We note that E. Witten's [34] proof implies this theorem under the (topological) assumption that the manifold is spin. The n-dimensional proof is given in [2,15].

Finally in the last section we discuss weak solutions of the Yamabe equation on S^n with prescribed singular set. We motivate this through the example of §2 which gives the solutions with two singular points. We also relate weak solutions to the geometry of locally conformally flat manifolds describing some of the results of [29]. Lastly we give a brief account of the author's existence theorem [24] for weak solutions with prescribed singular set.

1 The variational problem

Let M be a smooth n-dimensional compact manifold without boundary. For any smooth Riemannian metric g on M we let w_g denote the volume form of g; thus if $x^1, \ldots, x^n$ are local coordinates on M we have

$$g = \sum_{i,j=1}^{n} g_{ij}(x)\, dx^i dx^j\,, \qquad w_g = \sqrt{\det(g_{ij})}\, dx^1 \wedge \ldots \wedge dx^n\,.$$

Let $\mathcal{M}$ denote the space of all smooth Riemannian metrics on M, and let $\mathcal{M}_1$ denote the subset of $\mathcal{M}$ consisting of those metrics of total volume one; that is,

$$\mathrm{Vol}(g) = \int_M dw_g = 1\,.$$

Let $\mathrm{Riem}(g)$, $\mathrm{Ric}(g)$, $R(g)$ denote the Riemann curvature tensor, the Ricci tensor, and the scalar curvature respectively. In local coordinates we have

$$\mathrm{Riem}(g) = \sum_{i,j,k,\ell} R_{ijk\ell}(dx^i \wedge dx^j) \otimes (dx^k \wedge dx^\ell)$$

$$\mathrm{Ric}(g) = \sum_{i,j} R_{ij} dx^i dx^j\,, \qquad R_{ij} = \sum_{k,\ell} g^{k\ell} R_{ikj\ell}$$

$$R(g) = \sum_{i,j} g^{ij} R_{ij}\,.$$

The (elliptic) Einstein equations then express the condition that the trace-free part of the Ricci tensor vanishes, that is

$$\mathrm{Ric}(g) = \frac{1}{n} R(g)\, g\,. \tag{1.1}$$

The contracted second Bianchi identity implies

$$\sum_{j,k}^{n} g^{jk}(R_{ij} - \frac{1}{2} R(g)\, g_{ij})_{;k} = 0\,, \quad i = 1, \ldots, n$$

where the semi colon denotes the covariant derivative of a tensor with respect to the Levi Civita connection of g. Thus for $n \geq 3$ we see that (1.1) implies

$$R(g) \equiv \mathcal{R}(g) \tag{1.2}$$

where $\mathcal{R}(g) = \mathrm{Vol}(g)^{-1} \int_M R(g) d\omega_g$. It was shown by Hilbert that equation (1.1) arises as the Euler–Lagrange equations for the functional $\mathcal{R}(g)$ on the space $\mathcal{M}_1$. This may seem surprising since (1.1) is a second order equation for g while the integrand $R(g)$ of $\mathcal{R}(g)$ also involves second derivatives of g. To see that this is correct, let $g \in \mathcal{M}_1$ and let h be any smooth symmetric tensor of type $(0,2)$ on $\mathcal{M}$. We then set $g(t) = g + th$ for $t \in (-\varepsilon, \varepsilon)$, and this gives us a family of Riemannian metrics. The normalized family $\overline{g}(t) = V(t)^{-2/n} g(t)$,

$V(t) = \mathrm{Vol}(g+th)$ is then a path in $\mathcal{M}_1$. We have the formulae

$$R_{ij} = \sum_k \left\{ \Gamma^k_{ij,k} - \Gamma^k_{ki,j} + \sum_\ell (\Gamma^k_{k\ell}\Gamma^\ell_{ji} - \Gamma^k_{j\ell}\Gamma^\ell_{ki}) \right\}$$

$$\Gamma^k_{ij} = \frac{1}{2}\sum_\ell g^{k\ell}(g_{i\ell,j} + g_{j\ell,i} - g_{ij,\ell})$$

where the comma denotes the partial derivative in a local coordinate system. Using an "upper dot" to denote the derivative with respect to t, we have

$$\begin{aligned} \dot{R}_{ij} &= \sum_k (\dot{\Gamma}^k_{ij;k} - \dot{\Gamma}^k_{ki;j}) \\ \dot{\Gamma}^k_{ij} &= \frac{1}{2}\sum g^{k\ell}(h_{i\ell;j} + h_{j\ell;i} - h_{ij;\ell}). \end{aligned} \tag{1.3}$$

Therefore we find that $\dot{R}$ has the expression

$$\dot{R} = -\sum_{i,j} h^{ij}R_{ij} + \text{ divergence terms}$$

where $h^{ij} = \sum_{k,\ell} g^{ik}g^{j\ell}h_{k\ell}$. Upon integration we find

$$\begin{aligned} \frac{d}{dt}\int_M R(g(t))\,d\omega_{g(t)} &= -\int_M \langle h, \mathrm{Ric}\,(g(t))\rangle_{g(t)} d\omega_{g(t)} + \frac{1}{2}\int_M R(g(t))\mathrm{Tr}_{g(t)}(h)\,d\omega_{g(t)} \\ &= -\int_M \langle h, \mathrm{Ric}\,(g(t)) - \frac{1}{2}R(g(t))\,g(t)\rangle_{g(t)} d\omega_{g(t)} \end{aligned}$$

where we have used Stoke's theorem together with the formulas

$$\dot{\omega}_{g(t)} = \frac{1}{2}\mathrm{Tr}_{g(t)}(h)\omega_{g(t)},$$

$$\mathrm{Tr}_{g(t)}(h) = \sum_{i,j} g(t)^{ij}h_{ij}.$$

Now we have $\mathcal{R}(\overline{g}(t)) = V(t)^{(2-n)/n}\int_M R(g(t))d\omega_{g(t)}$, and hence we find

$$\frac{d}{dt}\mathcal{R}(\overline{g}(t)) = -V(t)^{(2-n)/n}\int_M \langle h, F(g(t))\rangle_{g(t)} d\omega_{g(t)} \quad \text{where}$$

$$F(g) = \mathrm{Ric}\,(g) - \frac{1}{2}R(g)g + \frac{n-2}{2n}\mathcal{R}(g)g.$$

To derive this expression we have used, in addition to our computation above, the formula $\dot{V}(t) = \frac{1}{2}\int_M \langle h, g(t)\rangle_{g(t)} d\omega_{g(t)}$. Therefore, if g is a critical point for $\mathcal{R}(\cdot)$ on $\mathcal{M}_1$ we find, setting $t = 0$, that $F(g) \equiv 0$. In particular, it follows that the trace-free part of $\mathrm{Ric}\,(g)$ vanishes and hence (1.1) holds.

Now suppose g is a solution of (1.1) so that in particular $F(g) \equiv 0$. We compute the second variation of $\mathcal{R}(\cdot)$ at g. We have

$$\left.\frac{d^2}{dt^2}\mathcal{R}(\overline{g}(t))\right|_{t=0} = -\int_M \langle h, \mathcal{L}h\rangle_g d\omega_g \tag{1.4}$$

where $\mathcal{L}h = \dot{F}(g(0))$. Thus $\mathcal{L}$ is a linear operator on symmetric (0,2) tensors given by

$$\mathcal{L}h = \dot{\mathrm{Ric}}(g) - \frac{1}{2}\dot{R}g - \frac{1}{n}Rh \tag{1.5}$$

which $\dot{\mathrm{Ric}}(g)$, $\dot{R}$ may be computed from (1.3) and we have used (1.2) and the fact that $\dot{\mathcal{R}}(g) = 0$. We write the space of symmetric (0,2) tensors as a sum of three subspaces S_0, S_1, S_2 where S_0 denotes those h which may be written $h = L_{\mathbf{X}}g$ (Lie derivative) for some vector field $\mathbf{X}$ on M, that is,

$$h_{ij} = \mathbf{X}_{i;j} + \mathbf{X}_{j;i}\,.$$

(The fact that this decomposition of smooth (0,2) tensors is valid is shown in [8].) The subspace S_1 denotes the pure trace tensors, that is, the h of the form $h = \eta g$ where η is a smooth function on M. Finally S_2 denotes those h which are orthogonal to both S_0 and S_1, that is, those h satisfying $\mathrm{Tr}_g(h) = 0$ and $\sum_{j,k} g^{jk}h_{ij;k} = 0$. Tensors $h \in S_2$ are referred to as transverse traceless tensors. Note that the subspace S_0 consists of those infinitesimal deformations of g which arise from diffeomorphisms of M. It follows that if $\mathbf{X}$ is a vector field on M and $\Phi_t : M \to M$ is the one-parameter group of diffeomorphisms generated by $\mathbf{X}$, then we have for each $t \in \mathbf{R}$, $F(\Phi_t^*) = 0$. Differentiating and setting $t = 0$ we have $\mathcal{L}h = 0$ where $h = L_{\mathbf{X}}g$. Thus $\mathcal{L} \equiv 0$ on S_0. We now compute $\mathcal{L}h$ for $h \in S_1$. Suppose $h = \eta g$ where η is a smooth function. We then have from (1.3), (1.5)

$$\mathcal{L}h = \frac{n-2}{2}\left((\Delta\eta + \frac{1}{n}R\eta)g - \mathrm{Hess}\,(\eta)\right) \tag{1.6}$$

where $\mathrm{Hess}\,(\eta) = \sum_{i,j}\eta_{;ij}dx^i dx^j$ is the Hessian of η. Now we have $\mathrm{Hess}\,(\eta) \in S_0$, so we see that $S_0 + S_1$ is invariant under $\mathcal{L}$.

Next we show that $\mathcal{L}$ is a self-adjoint operator. This may be seen from the variational definition of $\mathcal{L}$ by considering two symmetric (0,2) tensors h, k and the two parameter variation $g(t,s) = g + th + sk$. Let $\overline{g} = V(t,s)^{-2/n}g(t,s)$ be the normalized variation. We then have from above

$$\frac{\partial \mathcal{R}(\overline{g}(t,s))}{\partial t} = -V(t,s)^{(2-n)/n}\int_M \langle h, F(g(t,s))\rangle_{g(t,s)} d\omega_{g(t,s)}\,.$$

Differentiating in s and setting $t = s = 0$ we have

$$\left.\frac{\partial^2 \mathcal{R}(\overline{g}(t,s))}{\partial s \partial t}\right|_{t=s=0} = -\int_M \langle h, \mathcal{L}k\rangle_g d\omega_g\,.$$

Reversing the order of differentiation for the smooth function $\mathcal{R}(\overline{g}(t,s))$ of two variables we get

$$\int_M \langle h, \mathcal{L}k\rangle_g d\omega_g = \int_M \langle k, \mathcal{L}h\rangle_g d\omega_g$$

for all h, k. Thus $\mathcal{L}$ is self-adjoint.

Now since $S_0 + S_1$ is $\mathcal{L}$–invariant and $\mathcal{L}$ is self–adjoint it follows that $S_2 = (S_0 + S_1)^\perp \cap C^\infty$ is also $\mathcal{L}$–invariant. We compute $\mathcal{L}h$ for $h \in S_2$ using (1.3), (1.5)

$$(\mathcal{L}h)_{ij} = -\frac{1}{2}(\Delta h)_{ij} + \frac{1}{2}\sum_{k,\ell} g^{k\ell}(h_{ik;j\ell} + h_{jk;i\ell}) - \frac{1}{n} Rh_{ij}$$

where Δh is the trace Laplacian given by

$$(\Delta h)_{ij} = \sum_{k,\ell} g^{k\ell} h_{ij;k\ell}\,.$$

Using the transverse (divergence free) condition on h we may interchange covariant derivatives and write the second term above as a zero order term in h

$$\mathcal{L}h = -\frac{1}{2}\Delta h + K(h) \tag{1.7}$$

where $K(h)$ is the linear term

$$(K(h))_{ij} = -\sum_{k,\ell} R_{ikj\ell}h^{k\ell} + \frac{1}{2}\sum_k (R_{ik}h_j^k + R_{jk}h_i^k) - \frac{1}{n}Rh_{ij}\,.$$

An important qualitative feature of the variational problem is apparent from (1.6) and (1.7), namely that a critical metric g tends to minimize $\mathcal{R}$ among those metrics conformally equivalent to g and to maximize $\mathcal{R}$ among metrics transversely related to g. In fact, for $h = \eta g \in S_1$, we denote by $\mathcal{L}_1$ the second variation operator on the conformal class of g. Thus $\mathcal{L}_1$ is the operator $\mathcal{L}$ followed by projection into S_1. Precisely $\mathcal{L}_1$ is the scalar operator

$$\mathcal{L}_1\eta = \frac{(n-1)(n-2)}{2n}(\Delta\eta + \frac{1}{n-1}R\eta)\,. \tag{1.8}$$

Thus if we consider the restricted variational problem for the functional $\mathcal{R}(\cdot)$ on the conformal class $[g]$ of g we have for $h = \eta g$

$$\frac{d^2}{dt^2}\mathcal{R}(\overline{g}(t))\bigg|_{t=0} = -n\int_M \eta\mathcal{L}_1\eta\, d\omega_g\,.$$

Now the operator $-\mathcal{L}_1$ has eigenvalues tending to $+\infty$, and hence the metric g locally minimizes $\mathcal{R}$ in $\mathcal{M}_1 \cap [g]$ modulo a finite dimensional space of variations (finite Morse index). On the other hand, from (1.7) we see that the operator $-\mathcal{L}$ on S_2 has eigenvalues tending to $-\infty$ so that $\mathcal{R}(\cdot)$ is locally maximized among variations from S_2 modulo a finite dimensional subspace.

This dichotomy for the linearized operator $\mathcal{L}$ suggests the following global procedure for finding critical points of $\mathcal{R}(\cdot)$ on $\mathcal{M}_1$. For any $g_0 \in \mathcal{M}_1$, let $[g_0]$ denote the conformal class of g_0, that is,

$$[g_0] = \{g \in \mathcal{M} : \; g = e^{2v}g_0 \text{ for some } v \in C^\infty(M)\}\,.$$

Let $[g_0]_1 = \mathcal{M}_1 \cap [g_0]$, and define $I(g_0)$ by

$$I(g_0) = \inf\{\mathcal{R}(g) : \; g \in [g_0]_1\}\,.$$

If $g \in [g_0]_1$ realizes the infimum, then we see from above that the Euler–Lagrange equation satisfied by g is $\mathrm{Tr}_g(F(g)) \equiv 0$, that is, equation (1.2) holds. If we write $g = u^{4/(n-2)} g_0$ where u is a positive smooth function then we have the formula

$$R(g) = -c(n)^{-1} u^{-(n+2)/(n-2)} L_0 u$$

where $c(n) = \frac{n-2}{4(n-1)}$ and L_0 is the "conformal Laplacian" for the metric g_0

$$L_0 u = \Delta_{g_0} u - c(n) R(g_0) u \, .$$

Thus our functional $\mathcal{R}(\cdot)$ becomes $\mathcal{R}(g) = c(n)^{-1} E(u)$ where

$$E(u) = \int_M [|\nabla_{g_0} u|^2 + c(n) R(g_0) u^2] d\omega_{g_0} \, .$$

The volume constraint on g then becomes $\int_M u^{2n/(n-2)} d\omega_{g_0} = 1$. The equation (1.2) may then be written

$$L_0 u + c(n) \mathcal{R}(g) u^{(n+2)/(n-2)} = 0 \, . \tag{1.9}$$

Since

$$E(u) \geq \lambda_0(L_0) \int_M u^2 d\omega_{g_0} \geq \min\{0, \lambda_0(L_0)\} \, ,$$

where $\lambda_0(L_0)$ denotes the lowest eigenvalue of L_0, we see that $I(g_0) > -\infty$ for any g_0. We then define $\sigma(M)$ to be the supremum of $I(g_0)$ over all $g_0 \in \mathcal{M}_1$,

$$\sigma(M) = \sup\{(g_0) : \; g_0 \in \mathcal{M}_1\} \, .$$

If we consider constant curvature metrics g_0 on S^n normalized to have volume one, then we have $\mathcal{R}(g_0) = n(n-1)\mathrm{Vol}\,(S^n(1))^{2/n}$ where $S^n(1)$ denotes the sphere of radius 1. The following lemma tells us that the standard metric on S^n in fact realizes $\sigma(S^n)$ and provides an upper bound for $\sigma(M)$ for any n-dimensional manifold M.

Lemma 1.1. *We have $\sigma(S^n) = n(n-1)\mathrm{Vol}\,(S^n(1))^{2/n}$, and for any n-manifold M we have $\sigma(M) \leq \sigma(S^n)$.*.

Proof: Let $g_0 \in \mathcal{M}_1$ be a metric on M. We may show that $I(g_0) \leq n(n-1)\mathrm{Vol}\,(S^n)^{2/n}$ by constructing a metric $g \in [g]$, which is a concentrated spherical metric near a point of M. We omit the details and refer the reader to [1,15,23].

Now let g_0 be a constant curvature, unit volume metric on S^n. The fact that $\mathcal{R}(g_0) = I(g_0)$ follows from a symmetrization argument ([21,31]) or from the existence theory together with a uniqueness theorem of M. Obata (see later discussion) as in [15]. Combining these two facts we see that $\sigma(S^n) = \mathcal{R}(g_0) \geq \sigma(M)$ for any n-manifold M. This completes the proof of Lemma 1.1.

If we have a metric $g \in \mathcal{M}_1$ which realizes $\sigma(M)$, that is, $\mathcal{R}(g) = I(g) = \sigma(M)$, we should hope that g is Einstein. This is generally true if $\sigma(M) \leq 0$ but is not clear for $\sigma(M) > 0$.

To see this for $\sigma(M) \leq 0$, we use the fact that for $I(g_0) \leq 0$ there is a unique solution of (1.9). The existence follows from [32] and uniqueness from the maximum principle. Thus if h is any trace-free (0,2) tensor, we consider the deformed metric $g^{(t)} = g + th$. There is then a unique function $v^{(t)} > 0$ such that $(v^{(t)})^{4/(n-2)} g^{(t)}$ has constant scalar curvature equal to $I(g^{(t)})$ since $I(g^{(t)}) \leq \sigma(M) \leq 0$. The family $v^{(t)}$ is smooth as a function of t (see [14]), so we have $\frac{d}{dt} I(g^{(t)}) = 0$ at $t = 0$, and this tells us that the trace-free Ricci tensor of g vanishes and g is Einstein.

We now discuss properties of $\sigma(M)$ and various uniqueness theorems.

Lemma 1.2. *Let M be a smooth, closed n-dimensional manifold. The invariant $\sigma(M)$ is positive if and only if M admits a metric of positive scalar curvature.*

Proof: If $\sigma(M) > 0$, then by definition there is a metric $g_0 \in \mathcal{M}_1$ with $I(g_0) > 0$. This implies that $\lambda_0(L_0) > 0$, and hence the lowest eigenfunction u_0, which may be taken to be positive, satisfies $L_0 u_0 < 0$. Thus the metric $u_0^{4/(n-2)} g_0$ has positive scalar curvature.

Conversely, if $g_0 \in \mathcal{M}_1$ has positive scalar curvature, then $I(g_0) > 0$ (see [1]) and hence $\sigma(M) > 0$. This completes the proof of Lemma 1.2.

Since many topological obstructions are known for manifolds to admit metrics of positive scalar curvature (see [13,28]), Lemma 1.2 indicates that the invariant $\sigma(M)$ is quite nontrivial. We prove the following uniqueness theorem for constant curvature metrics.

Proposition 1.3.

1. *Let $M = S^n$. Any metric $g \in \mathcal{M}_1$ which satisfies $\mathcal{R}(g) = I(g) = \sigma(g)$ has constant positive sectional curvature.*

2. *Suppose that M admits a flat metric. Any metric $g \in \mathcal{M}_1$ satisfying $\mathcal{R}(g) = I(g) = \sigma(M)$ is a flat metric. In particular, $\sigma(M) = 0$ and any flat metric $g \in \mathcal{M}_1$ satisfies $\mathcal{R}(g) = I(g) = \sigma(M)$.*

Proof: The proof of the first statement is a consequence of the work ([1,23]) on the Yamabe problem which shows that $I(g_0) < \sigma(S^n)$ for any $g_0 \in \mathcal{M}_1$ unless g_0 has constant curvature ($M = S^n$). Statement 2 follows from [13,28] where it is shown that a flat manifold does not admit a metric of positive scalar curvature (i.e. $\sigma(M) \leq 0$), and any scalar flat metric on M is flat. This completes the proof of Proposition 1.3.

There are two obvious uniqueness questions left unresolved for metrics of constant curvature. The first is whether the constant positive curvature metrics g on non-simply connected manifolds achieve the same characterization as the standard metrics on S^n, i.e. $\mathcal{R}(g) = I(g) = \sigma(M)$. The second question is whether a hyperbolic metric g on M can be characterized similarly. We conjecture that the answer is yes to these questions.

As a final topic in this section we discuss the uniqueness theorem of Obata [18] and its relevance to our variational problem. Let $g_0 \in \mathcal{M}$, and let $g = u^{4/(n-2)}g_0$ where u is a smooth positive function. Let T_0, T denote the trace-free part of the Ricci tensors of g_0, g respectively. We then have the formula

$$T = T_0 + (n-2)u^{2/(n-2)}\left[\operatorname{Hess}(u^{-2/(n-2)}) - \frac{1}{n}\Delta(u^{-2/(n-2)})g\right] \tag{1.10}$$

which follows from direct computation (see [4]). In (1.10) the Hessian and Laplacian are with respect to g_0. Assume g_0 has constant scalar curvature. We then have by the contracted Bianchi identity $\sum_{j,k} g_0^{jk}(T_0)_{ij;k} = 0$ for $i = 1, \ldots, n$. It follows then from Stoke's theorem

$$\int_M \left\langle T_0, \operatorname{Hess}(u^{-2/(n-2)})\right\rangle_{g_0} d\omega_{g_0} = 0\,.$$

Therefore, we multiply (1.10) by $u^{-2/(n-2)}$ and integrate its inner product with T_0 to get

$$\int_M u^{-2/(n-2)}\langle T, T_0\rangle_{g_0} d\omega_{g_0} = \int_M u^{-2/(n-2)}\|T_0\|_{g_0}^2 d\omega_{g_0}\,. \tag{1.11}$$

Combining (1.11) with the Schwarz inequality we see that for *any* constant scalar curvature metric g_0 and for any $g = u^{4/(n-2)}g_0$ we have

$$\int_M u^{-2/(n-2)}\|T_0\|_{g_0}^2 d\omega_{g_0} \le \int_M u^{-2/(n-2)}\|T\|_{g_0}^2 d\omega_{g_0}\,. \tag{1.12}$$

In particular, if g were Einstein then g_0 would necessarily also be Einstein.

Proposition 1.4. *For an Einstein metric g (unit volume) on M we necessarily have $\mathcal{R}(g) = I(g)$. Moreover, any constant scalar curvature metric $g_0 \in [g]_1$ is Einstein. We then have $g_0 = g$ unless (M, g) is isometric to a round S^n in which case g_0 is a constant curvature metric on S^n which is pointwise conformal to g.*

The main step in the proof of this result is (1.12) which shows that g_0 is Einstein if it has constant scalar curvature. The analysis of conformally related Einstein metrics on a closed manifold is fairly straightforward (again based on (1.10)) and we omit the details referring the reader to Obata [18] for the complete proof.

A consequence of Proposition 1.4 is that any critical point $g \in \mathcal{M}_1$ of $\mathcal{R}(\cdot)$ automatically minimizes in its conformal class and hence has conformal Morse index zero.

We also observe that for $n = 3$ inequality (1.12) gives a strong a priori estimate on solutions of the Yamabe equation (1.9). To agree with our earlier notation we let $g_0 \in \mathcal{M}_1$ be a fixed metric and let $g = u^{4/(n-2)}g_0$ have constant scalar curvature. Inequality (1.12) then says for $n = 3$ (note that u of (1.12) becomes u^{-1})

$$\int_M u^2\|T\|_g^2 d\omega_g \le \int_M u^2\|T_0\|_g^2 d\omega_g\,.$$

Since $n = 3$ we also have

$$\int_M \|T_0\|_{g_0}^2 d\omega_g = \int_M u^2\|T_0\|_g^2\, d\omega_g\,.$$

Therefore we have

$$\begin{aligned}\left(\int_M (\|T\|_g)^{3/2} d\omega_g\right)^{2/3} &\le \left(\int_M u^{-6} d\omega_g\right)^{1/6} \left(\int_M u^2 \|T\|_g^2 d\omega_g\right)^{1/2} \\ &\le \left(\int_M \|T_0\|_{g_0}^2 d\omega_{g_0}\right)^{1/2}\end{aligned} \tag{1.13}$$

since $u^{-6} d\omega_g = d\omega_{g_0}$, and we assume $\mathrm{Vol}\,(g_0) = 1$. It follows that the quantity $\int_M (\|T\|_g)^{3/2} d\omega_g$ is a priori bounded (depending only on the background g_0, hence the conformal class) for any metric $g \in [g_0]$ of constant scalar curvature. Note that $\int_M (\|T\|_g)^{3/2} d\omega_g$ is a dimensionless quantity for $n = 3$.

2 The Yamabe problem

In this section we discuss solvability of (1.2), or equivalently (1.9). From the previous section we know that (1.2), (1.9) is the Euler–Lagrange equation for the functional $\mathcal{R}(\cdot)$ on $[g_0]_1$. An approach to producing solutions of this equation would be to construct a minimizer; that is, a metric $g \in [g_0]_1$ such that $\mathcal{R}(g) = I(g_0)$. This approach has been successful as we will outline here.

Historically this problem was studied by H. Yamabe [35] in the early sixties, and was claimed to have been solved in [35]. During the sixties there was substantial development in partial differential equations, and nonlinear problems were being understood more deeply. In particular it was realized [20] that, in many situations, equations such as (1.9) do not have positive solutions. In light of these developments, N. Trudinger re-examined Yamabe's paper and discovered that it contained a serious error. In [32] Trudinger developed analytic machinery relevant to (1.9) and showed that a solution (in fact a minimizer) exists if $I(g_0) \leq 0$ (or if $I(g_0)$ is not too positive). He also proved regularity of $W^{1,2}$ weak solutions of (1.9). This left open the general case with $I(g_0) > 0$. The fact that this case is subtle is apparent from the example of (S^n, g_0) where $g_0 \in \mathcal{M}_1$ has constant sectional curvature. In this case, g_0 is itself a solution of (1.2) but is by no means the only solution in $[g_0]_1$. In fact, given any conformal transformation $F : S^n \to S^n$ we have $F^*(g_0) \in [g_0]_1$ is another solution of (1.2). Thus if we take a divergent sequence of conformal transformations F_i (such as dilations $F_i(x) = i \cdot x$ in stereographic coordinates) we get a divergent sequence of minima for the Yamabe problem on (S^n, g_0). In particular, one cannot obtain uniform estimates on solutions such as would be required to prove existence by usual analytic methods. It follows that any method which produces solutions "with bounds" must distinguish (S^n, g_0) from the conformal class one considers. In 1976, T. Aubin [1] proved a general existence result in the positive case. He showed that if $n \geq 6$ and g_0 is not locally conformally flat then (1.2) has a solution (in fact, a minimizer) $g \in [g_0]_1$. A metric g_0 is said to be locally conformally flat if in a neighborhood of any point of M, there exists local coordinate $x^1, \ldots, x^n$ such that g_0 is given by

$$g_0 = \lambda^2(x) \sum_{i=1}^{n} (dx^i)^2$$

for a locally defined positive function $\lambda(x)$. Alternatively, a metric g_0 is locally conformally flat if any point $p_0 \in M$ has a neighborhood ϑ such that (ϑ, g_0) is conformally equivalent to a subdomain of the standard sphere. In particular, the assumption that g_0 be not locally conformally flat should be viewed as requiring (M, g_0) to be far from the standard sphere (which we've seen is a bad case). By a purely local computation Aubin showed that a manifold (M, g_0) with $n \geq 6$ and g_0 not l.c.f. satisfies $I(g_0) < \sigma(S^n)$ and thus one can derive the necessary estimates to construct a minimizer. We refer the reader to [15] for details and merely describe developments here in a general way. Because Aubin's argument is purely local, there was no chance that it could work for a locally conformally flat metric, and all attempts to weaken the dimensional restriction ($n \geq 6$) have failed. In 1984 (see [23]) we developed a new global attack on the problem and succeeded in solving (1.9) (again producing a minimizer) for $n = 3, 4, 5$ and for locally conformally flat metrics. We present here the general idea and refer the reader to [15] for details. (In the next section we present an a priori estimate for solutions of (1.2) which are not necessarily minimizers.) The critical metrics one must consider in the Yamabe problem are those which are concentrated near a point p_0 of M, and are very small away from the point. If g denotes such a metric, then we may choose a point $p \neq p_0$, and rescale g by multiplication by a large constant so that g agrees with our background metric g_0 at p. If we imagine a sequence of metrics $\{g_i\} \in \mathcal{M}_1$ which concentrate near p_0 and tend to zero at p, then by rescaling we get a sequence $\{\overline{g}_i\}$ which are uniformly controlled near p. If the scalar curvatures of the g_i were bounded, then the scalar curvatures of $\overline{g}_i$ tend to zero, and we expect the $\overline{g}_i$ to converge to a metric $\overline{g}$ of zero scalar curvature with $\overline{g}$ being a complete metric on $M - \{p_0\}$. (We rigorously carry out this type of rescaling in the next section.) If we write $\overline{g} = G^{4/(n-2)} g_0$ as a function times our background g_0, then G satisfies $L_0 G = 0$ on $M - \{p_0\}$, and $G > 0$. Thus G must be a (multiple of) the fundamental solution of L_0 with pole at p_0. Near p_0, the function G has the behavior $G(x) = |x|^{2-n} + \alpha(x)$ where α has a milder singularity at $x = 0$ than $|x|^{2-n}$. Thus near p_0, the metric $\overline{g}$ approximates $|x|^{-4} \sum_i dx_i^2$ which is simply the metric $\sum dy_i^2$ on $\mathbf{R}^n$ written in the inverted coordinates $y = |x|^{-2} x$. Thus $(M - \{p_0\}, \overline{g})$ is scalar flat and asymptotically flat. In such a situation (in certain cases) there is a number which can be attached to $\overline{g}$ which is referred to as total energy. The reason for this name is that for $n = 3$, asymptotically flat manifolds arise as initial data for asymptotically flat spacetimes which model finite isolated gravitating systems in general relativity. The scalar curvature assumption corresponds to (a special case of) the physical assumption that the local energy density of the matter fields be nonnegative. The total energy of a system measures the deviation of $\overline{g}$ from the Euclidean metric at infinity, and "positive energy" theorems assert that the total energy is strictly positive unless $(M - \{p_0\}, \overline{g})$ is isometric to $\mathbf{R}^n$. In [23], it is shown that if g_0 is locally conformally flat or if $n = 3, 4, 5$ the energy term can be used to show that $I(g_0) < \sigma(S^n)$ unless (M, g_0) is conformally equivalent to the standard S^n. This implies existence of a minimizer for the Yamabe problem with appropriate estimate. In §4 we discuss the positive energy theorems which are needed for the Yamabe problem.

For a compact, closed manifold M, let $g_0 \in \mathcal{M}_1$, and let $\mathcal{F}$ be given by

$$\mathcal{F} = \{u : \ u^{4/(n-2)} g_0 \in \mathcal{M}_1, \ \mathcal{R}(u^{4/(n-2)} g_0) = I(g_0)\} .$$

Thus $\mathcal{F}$ is the set of solutions of (1.9) which arise as minimizers for the Yamabe problem. The following compactness theorem is a standard consequence (see [23]) of the inequality $I(g_0) < \sigma(S^n)$.

Proposition 2.1. *Suppose (M, g_0) is not conformally equivalent to the standard sphere. The* *set $\mathcal{F}$ is a nonempty compact subset of $C^2(M)$, the set of twice continuously differentiable functions on M with the usual C^2 norm.*

As we have observed, the above result is false for the standard sphere because the conformal group of S^n is noncompact. We give a geometric corollary which says that any manifold except the standard sphere has a compact conformal group. This result is a theorem of J. Lelong-Ferrand [16].

Corollary 2.2. *If (M, g_0) is not conformally equivalent to the standard sphere, then the group of conformal automorphisms of (M, g_0) is compact.*

Proof: Let $\mathcal{D}$ be the group of conformal diffeomorphisms of (M, g_0). It suffices to show that $\mathcal{D}$ is compact in the C^0 topology. The main point is that $\mathcal{D}$ acts on the set $\mathcal{F}$ by pullback; that is, given $F \in \mathcal{D}$, $u \in \mathcal{F}$ we have $F^*(u^{4/(n-2)}g_0) = (u_F)^{4/(n-2)}g_0$, $u_F = |F'|^{(n-2)/2}u \circ F \in \mathcal{F}$. Here we write $F^*g_0 = |F'|^2 g_0$ so that $|F'|$ is a function which measures the stretch factor of F measured with respect to g_0. Thus the compactness of $\mathcal{F}$ implies that $u_F \leq c$ for all $F \in \mathcal{D}$, and hence $|F'|$ is uniformly bounded for all $F \in \mathcal{D}$. Therefore, by the Arzela-Ascoli theorem, $\mathcal{D}$ is a compact subset of $C^0(M, M)$. This completes the proof of Corollary 2.2.

There are very few (conformal) manifolds on which one can analyze all solutions of (1.2). Besides the standard sphere, where Obata's theorem tells us that all solutions are minimizing and have constant sectional curvature, the product metrics on $S^1 \times S^{n-1}$ seem to be the only manifolds where all solutions can be analyzed. In particular, on $S^1 \times S^{n-1}$ we see many solutions of (1.2) which are *not* minimizing, and we see situations where the most symmetric solutions are not the minima. For convenience of notation, we dispense with the volume constraint and normalize solutions of (1.2) so that their scalar curvature is equal to $n(n-1)$, the scalar curvature of the unit n-sphere. Equation (1.9) then becomes

$$L_0 u + \frac{n(n-2)}{4} u^{(n+2)/(n-2)} = 0 . \tag{2.1}$$

We analyze $S^1 \times S^{n-1}$ by looking for solutions on the universal covering space $\mathbf{R} \times S^{n-1}$, and we choose S^{n-1} to have unit radius. If we consider the n-sphere to be $\mathbf{R}^n \cup \{\infty\}$ where the coordinates $x \in \mathbf{R}^n$ arise from stereographic projection, then the manifold $\mathbf{R} \times S^{n-1}$ is conformally equivalent to $S^n - \{0, \infty\} = \mathbf{R}^n - \{0\}$. The conformal diffeomorphism is given explicitly by sending the point $x \in \mathbf{R}^n - \{0\}$ to the point $(\log|x|, x/|x|) \in \mathbf{R} \times S^{n-1}$. Thus the analysis of solutions of (1.9) on $\mathbf{R} \times S^{n-1}$ is completely equivalent to the analysis of solutions of (1.9) on $\mathbf{R}^n - \{0\}$. An important method was introduced into the subject by Gidas, Ni, and Nirenberg [9] which enables one to show that, under suitable conditions, arbitrary solutions of (1.9) have a maximal amount of symmetry. For solutions on $S^n - \{0, \infty\}$ it has been shown by Caffarelli, Gidas, Spruck [5] that any solution of (1.9) which is singular at either 0 or ∞ is necessarily singular at both 0 and ∞, and such a solution is a radial function, that is, a function of $|x|$. We are interested in complete metrics on $\mathbf{R} \times S^{n-1}$ and hence we want solutions singular at both 0 and ∞. We will write (2.1) with respect to the product metric $g_0 = dt^2 + d\xi^2$ on $\mathbf{R} \times S^{n-1}$ where (t, ξ) denote coordinates on $\mathbf{R} \times S^{n-1}$, and $d\xi^2$ is used to denote the metric on the unit S^{n-1}. We then have $R(g_0) = (n-1)(n-2)$, and for a function $u(t)$ (which any global solution will be from the above discussion), equation (2.1) becomes

$$\frac{d^2u}{dt^2} - \frac{(n-2)^2}{4} u + \frac{n(n-2)}{4} u^{(n+2)/(n-2)} = 0. \tag{2.2}$$

We are interested in positive solutions of (2.2) defined on all of $\mathbf{R}$. There are two obvious nonzero solutions of (2.2). The first is the constant solution

$$u(t) \equiv u_0 = \left(\frac{n-2}{n}\right)^{(n-2)/4} \tag{2.3}$$

Geometrically, $u_0^{4/(n-2)} g_0$ is that multiple of g_0 having scalar curvature $n(n-1)$. The second explicit solution is a solution of constant sectional curvature. The spherical metric g_1 on $\mathbf{R}^n$ takes the form $g_1 = 4(1+|x|^2)^{-2} \sum_i (dx^i)^2$. Writing this metric as a function times g_0 we get

$$g_1 = 4(|x| + |x|^{-1})^{-2} g_0 = (\cosh t)^{-2} g_0 .$$

Therefore the function $u_1(t)$ given by

$$u_1(t) = (\cosh t)^{-(n-2)/2} \tag{2.4}$$

is a solution of (2.2). Of course the metric g_1 is *not* a complete metric on $\mathbf{R} \times S^{n-1}$. We convert (2.2) to a first order system by setting $v = \frac{du}{dt}$, and defining the vector field $\mathbf{X}(u, v)$ in the uv-plane by

$$\mathbf{X}(u,v) = \left(v, \frac{(n-2)^2}{4} u - \frac{n(n-2)}{4} u^{(n+2)/(n-2)}\right) .$$

Equation (2.2) then becomes the autonomous system

$$\frac{d}{dt}(u,v) = \mathbf{X}(u,v) .$$

The vector field $\mathbf{X}$ has critical points at $(0,0)$ and $(u_0, 0)$. The linearized equation at $(0,0)$ is

$$\frac{du}{dt} = v , \quad \frac{dv}{dt} = \frac{(n-2)^2}{4} u$$

which has a saddle point at the origin. At $(u_0, 0)$ the linearized system becomes

$$\frac{du}{dt} = v , \quad \frac{dv}{dt} = (2-n)u$$

which has a proper node at the origin. The orbit corresponding to the solution $u_1(t)$ contains the point $(1,0)$, is symmetric under reflection in the u-axis, and approaches $(0,0)$ as t approaches both $+\infty$ and $-\infty$. Therefore, this orbit (together with $(0,0)$) bounds a region Ω, and the point $(u_0, 0)$ lies in Ω. Thus the region Ω is invariant under the flow, and it is elementary that any orbit on which u remains positive for all time must lie in $\overline{\Omega}$. We may parametrize the orbits in Ω by letting $\gamma_\alpha(t)$ denote the orbit with $\gamma_\alpha(0) = (\alpha, 0)$ where $\alpha \in [u_0, 1]$. Thus $\gamma_{u_0}(t) \equiv (u_0, 0)$, and $\gamma_1(t) = \left(u_1(t), \frac{du_1}{dt}(t)\right)$. For $\alpha \in (u_0, 1)$, there is a first positive time, which we denote $\frac{1}{2}T(\alpha)$, at which γ_α intersects the u-axis. We also see that if we denote the coordinates of $\gamma_\alpha(t)$ by $(u_\alpha(t), v_\alpha(t))$, then we have $\gamma_\alpha(-t) = (u_\alpha(t), -v_\alpha(t))$. Therefore it follows that $\gamma_\alpha(t)$ is periodic with period $T(\alpha)$. It should be true that $T(\alpha)$ is an increasing function of α, but we have not checked this. It is elementary that $\lim_{\alpha\uparrow 1} T(\alpha) = \infty$, and $\lim_{\alpha\downarrow u_0} T(\alpha) = (n-2)^{-1/2}2\pi$. The quantity $(n-2)^{-1/2}2\pi$ is the fundamental period of the linearized operator at u_0, which is $\frac{d^2}{dt^2} + (n-2)$.

We now summarize the consequences of the above discussion for solutions of (1.2) on $S^1 \times S^{n-1}$. We normalize the radius of S^{n-1} to be one, and let the length of S^1 be a parameter T, so our manifold is $S^1(T) \times S^{n-1}$. We take our background metric g_0 to be the product metric. We assume in this discussion that $T(\alpha)$ is increasing for $\alpha \in [u_0, 1]$, otherwise one can make the obvious modifications. There is a number $T_0 = (n-2)^{-1/2}2\pi$ such that for $T \le T_0$ the manifold $S^1(T) \times S^{n-1}$ has a unique solution for (2.1) hence for the Yamabe problem. This solution is a constant times g_0. For $T \in (T_0, 2T_0]$ equation (2.1) has two inequivalent solutions, the constant solution and also the solution with fundamental period T. Actually, since the solution with fundamental period T is not invariant under rotation about S^1 we actually have an S^1 parameter family of solutions. For $T \in (2T_0, 3T_0]$ we have 3 inequivalent solutions, the constant solution, two periods of the solution with fundamental period $T/2$, and the solution with fundamental period T. Again the last two lie in S^1 parameter families of solutions. Generally, we see that for $T \in ((k-1)T_0, kT_0]$ we have k inequivalent solutions given by the constant solution, together with i periods of a solution with fundamental period T/i for $i = 1, \ldots, k-1$. Each of these $(k-1)$ solutions lies in an S^1 parameter family of equivalent solutions. All of the solutions for $T > T_0$ are variationally unstable except the solutions with fundamental period T, and hence these solutions are minimizing for the Yamabe problem (after one normalizes the volume). The instability of the constant solution is elementary, and for a solution consisting of i periods of a solution with fundamental period T/i $(i \ge 2)$ we can use the following argument: Let $u(t)$ be such a solution. Then we have $u(t + T/i) = u(t)$, and hence $v(t) = \frac{du}{dt}$ has the property that $\{t \in S^1(T) : v(t) > 0\}$ consists of at least i disjoint intervals. On the other hand v satisfies the linearized equation

$$Lv = \frac{d^2v}{dt^2} + \left(\frac{n(n+2)}{4}u^{4/(n-2)} - \frac{(n-2)^2}{4}\right)v = 0\,.$$

It follows from Sturm–Liouville theory that there are at least i (≥ 2) eigenvalues of $-L$ which are less than zero. This implies instability for the constrained variational problem.

Since the solution with fundamental period T approaches u_1 as $T \to \infty$, we also see that

$$\lim_{T\to\infty} I(S^1(T) \times S^{n-1}) = \sigma(S^n)\,,$$

and in particular we have $\sigma(S^1 \times S^{n-1}) = \sigma(S^n)$ since we have exhibited a maximizing sequence of conformal classes of metrics on $S^1 \times S^{n-1}$. We see that $\sigma(S^1 \times S^{n-1})$ is not achieved by a smooth metric on $S^1 \times S^{n-1}$.

3 A priori estimates on nonminimal solutions

In this section we will derive estimates on metrics in a given conformal class which satisfy a generalization of equation (1.9). It will be essential for these estimates that (M, g_0) be conformally inequivalent to the standard sphere, as they are false on S^n. While analogues of these estimates hold in general, we restrict ourselves here to metrics g_0 which are locally conformally flat. This case contains the main ideas without as many technical complications as one encounters generally. We begin with a geometric Pohozaev-type identity which holds in exact form for a locally conformally flat metric g. Throughout this section we will assume that (M, g_0) is a locally conformally flat manifold and $g \in [g_0]$. Assume that $x^1, \ldots, x^n$ are local coordinates on M in which g takes the form $\lambda^{4/(n-2)}(x) \sum_i (dx^i)^2$. Let $r^2 = \sum_i (x^i)^2$ be the square of the Euclidean length of x, and let D_σ denote the open Euclidean ball centered at $x = 0$ of radius σ. The following identity holds

$$\int_{D_\sigma} r \frac{\partial R(g)}{\partial r} \, d\omega_g = \frac{2n}{n-2} \int_{\partial D_\sigma} T\left(r \frac{\partial}{\partial r}, \lambda^{-2/(n-2)} \frac{\partial}{\partial r}\right) d\Sigma_g \tag{3.1}$$

where $d\Sigma_g$ is surface measure on ∂D_σ determined by g and $T(\cdot, \cdot)$ is the trace-free Ricci tensor of g considered as a symmetric bilinear form on tangent vectors. The identity (3.1) reduces to the standard Pohozaev [20] identity for the function $\lambda(x)$. In this form it is derived in [24, Proposition 1.4] where the conformal Killing vector field is $\mathbf{X} = r\frac{\partial}{\partial r}$, the generator of dilations centered at 0 (locally defined). Suppose $g_0 = \lambda_0^{4/(n-2)}(x) \sum_i (dx^i)^2$ and $g = u^{4/(n-2)} g_0$ so that $\lambda = u\lambda_0$. We may rewrite (3.1)

$$\int_{D_\sigma} r \frac{\partial R(g)}{\partial r} (\lambda_0 u)^{2n/(n-2)} \, dx = \frac{2n}{n-2} \int_{\partial D_\sigma} \sigma^n (\lambda_0 u)^2 T\left(\frac{\partial}{\partial r}, \frac{\partial}{\partial r}\right) d\xi \tag{3.2}$$

where $d\xi$ denotes the volume measure on the unit $(n-1)$-sphere. Equation (1.10) gives us an expression for $T\left(\frac{\partial}{\partial r}, \frac{\partial}{\partial r}\right)$

$$T\left(\frac{\partial}{\partial r}, \frac{\partial}{\partial r}\right) = (n-2)(\lambda_0 u)^{2/(n-2)} \left[\frac{\partial^2}{\partial r^2}((\lambda_0 u)^{-2/(n-2)}) - \frac{1}{n} \Delta((\lambda_0 u)^{-2/(n-2)})\right] \tag{3.3}$$

where Δ denotes the Euclidean Laplace operator $\sum_i \frac{\partial^2}{(\partial x^i)^2}$.

A common method of attack on the existence of solutions of (1.9), which was in fact used by Yamabe, is to regularize the problem by lowering the exponent of the nonlinear term. Thus one introduces the equation

$$Lu + Ku^p = 0\,, \quad u > 0 \tag{3.4}$$

where K is a positive constant and $p \in (1, (n+2)/(n-2)]$. For $p < (n+2)/(n-2)$ it is standard to construct a nonzero solution which minimizes the associated constrained variational problem. More generally, the associated variational problem satisfies the Palais–Smale condition, and hence the methods of nonlinear functional analysis and the calculus of variations may be applied. We will derive uniform estimates on solutions of (3.4) which have bounded energy. In particular, these estimates imply that solutions of (3.4) converge in C^2 norm as $p \uparrow (n+2)/(n-2)$ to solutions of (1.9). We define for $\Lambda > 0$ a set of solutions $\mathcal{S}_\Lambda$

$$\mathcal{S}_\Lambda = \{u : u \text{ satisfies (3.4) for some } p \in \left(1, \frac{n+2}{n-2}\right], \; E(u) \le \Lambda, K \le \Lambda\}.$$

We will show that, if (M, g_0) is not conformally equivalent to S^n, then $\mathcal{S}_\Lambda$, is a compact subset of $C^2(M)$. We first state, without giving a detailed proof, a general weak compactness theorem for metrics $g \in [g_0]$ whose scalar curvatures are controlled. This type of result is at present well known to experts in several areas. An analogous theorem is proven by Sacks–Uhlenbeck [22] for harmonic maps in two variables, by Uhlenbeck [33] for Yang–Mills connections in four variables, and by several authors [7], [12], [17] in various contexts.

Proposition 3.1. *Let $\{u_i\}$ be a sequence of positive C^2 functions on M such that*

$$\{\mathrm{Vol}\,(u_i^{4/(n-2)}g_0)\}, \qquad \{R(u_i^{4/(n-2)}g_0)\}$$

are both uniformly bounded sequences. There is a subsequence $\{u_{i'}\}$ which converges weakly in $W^{1,2}(M)$ to a limit function u. The function u is C^1 on M, and there is a finite set of points $\{p_1, \ldots, p_k\}$ such that $\{u_{i'}\}$ converges in C^1 norm to u on compact subsets of $M - \{p_1, \ldots, p_k\}$.

Since our arguments will be geometric in nature, it will be convenient to estimate

$$R(u^{4/(n-2)}g_0)$$

for $u \in \mathcal{S}_\Lambda$. This can be done based on "subcritical" estimates.

Proposition 3.2. *Suppose $u \in \mathcal{S}_\Lambda$. There is a constant C depending only on g_0, Λ such that $\max|R(u^{4/(n-2)}g_0)| \le C$. Similarly all derivatives of $R(u^{4/(n-2)}g_0)$ with respect to g_0 can be bounded in terms of g_0, Λ.*

Proof: Let $\delta = (n+2)/(n-2) - p$ where u satisfies (3.4) with exponent p. If $\delta = 0$, then $R(u^{4/(n-2)}g_0) = c(n)^{-1}K$ and our result is trivial. Thus we assume $\delta > 0$, and we derive estimates on u keeping track of the δ-dependence. We first derive an upper bound on u by a scaling argument. Let $\overline{u} = \max\{u(p) : p \in M\}$ and let $\overline{p} \in M$ be a point with $u(\overline{p}) = \overline{u}$. Let $x^1, \ldots, x^n$ be coordinates centered at $\overline{p}$. Observe that for $a > 0$ the function $u_a(x)$ defined (locally) by $u_a(x) = a^{2/(p-1)}u(ax)$ satisfies the equation $L_a u_a + K u_a^p = 0$ where L_a is the operator

$$L_a v(x) = \frac{1}{\sqrt{\det g_0(ax)}} \sum_{i,j} \frac{\partial}{\partial x^i}\left(\sqrt{\det g_0(ax)}\, g_0^{ij}(ax)\frac{\partial v}{\partial x^j}\right) - c(n)a^2 R(g_0)(ax)v(x).$$

We choose a such that $u_a(0) = 1$, that is, we set $a = (\overline{u})^{-(p-1)/2}$. We assume $\overline{u}$ is large so that u_a is defined on the unit ball in $\mathbf{R}^n$. Since $x = 0$ is the maximum point of u_a in B_1, we have

$u_a \leq 1$ and standard elliptic estimates imply

$$u_a(0) \leq c\left(\int_{B_1} u_a^{2n/(n-2)} dx\right)^{(n-2)/(2n)}$$

Now from the definition of u_a we have after a change of variable,

$$\int_{B_1(0)} u_a^{2n/(n-2)} dx = a^{\frac{n}{p-1}\left(\frac{n+2}{n-2}-p\right)} \int_{B_a(0)} u^{2n/(n-2)} dx\,.$$

This then implies

$$1 = u_a(0) \leq c \cdot (\overline{u})^{-\frac{n-2}{4}\left(\frac{n+2}{n-2}-p\right)},$$

and hence we have

$$\max_M u \leq c_1^{1/\delta}, \quad \delta = \frac{n+2}{n-2} - p. \tag{3.5}$$

for a constant c_1.

We may derive a lower bound on u of a similar type by observing that $Lu \leq 0$, and so standard estimates (see [10]) give us

$$\min_M u \geq c \int_M u\, d\omega_{g_0}\,.$$

From (3.5) we have

$$\int_M u^{2n/(n-2)} d\omega_{g_0} \leq c_1^{\frac{n+2}{n-2}\cdot\frac{1}{\delta}} \int_M u\, d\omega_{g_0}\,.$$

The Sobolev inequality implies

$$\left(\int_M u^{2n/(n-2)} d\omega_{g_0}\right)^{(n-2)/n} \leq c\, E(u) = cK \int_M u^{p+1} d\omega_{g_0}\,.$$

Since $p+1 \leq 2n/(n-2)$ and $K \leq \Lambda$ we have

$$1 \leq c\Lambda \left(\int_M u^{2n/(n-2)} d\omega_{g_0}\right)^{\frac{n-2}{2n}(p-1)}.$$

Combining the above inequalities we get

$$\min_M u \geq c_2^{-1/\delta} \tag{3.6}$$

for a constant c_2. Rescaling as above with $a = \overline{u}^{-(p-1)/2}$ and with center any given point of M we get from elliptic theory $|\nabla u_a(0)| \leq c\, u_a(0) \leq c$ which implies in light of (3.5)

$$\max_M |\nabla_{g_0} u| \leq c_3^{1/\delta}. \tag{3.7}$$

Higher derivatives can be similarly estimated. To complete the proof we observe that

$$R(u^{4/(n-2)} g_0) = c(n)^{-1} K u^{-\delta}$$

from (3.4). Therefore (3.5), (3.6), (3.7) imply that $R(u^{4/(n-2)}g_0)$ and its first derivative with respect to g_0 are bounded. Higher derivatives of $R(u^{4/(n-2)}g_0)$ are similarly bounded, and we have completed the proof of Proposition 3.2.

We now prove the main result of this section.

Theorem 3.3. *Suppose (M, g_0) is not conformally equivalent to the standard n-sphere and g_0 is locally conformally flat. For any $\Lambda > 0$, the set S_Λ is a bounded subset of $C^3(M)$.*

Proof: We prove the theorem by contradiction. Suppose $\{u_i\}$ is a sequence in S_Λ with $\lim \|u_i\|_{C^3(M)} = \infty$. From Proposition 3.1 we may require the sequence u_i to converge weakly in $W^{1,2}(M)$ to a limit u, and uniformly on compact subsets of $M - \{P_1, \ldots, P_k\}$ for some collection of points $P_1, \ldots, P_k \in M$. The function u is smooth on M, and the sequence u_i converges in C^3 norm to u on compact subsets of $M - \{P_1, \ldots, P_k\}$ by elliptic estimates. If we can show that the sequence u_i converges uniformly on all of M, then we conclude that $\max u_i$ are bounded, and standard elliptic theory implies $\|u_i\|_{C^3(M)}$ are bounded contrary to assumption.

We divide the proof into two steps. We first show that u is nonzero. This is where we use the global hypothesis that (M, g_0) is not conformally S^n. Assume $u \equiv 0$, and choose a point $Q \in M$ different from $P_1, \ldots, P_k$. Let $\varepsilon_i = u_i(Q)$, so by assumption $\lim \varepsilon_i = 0$. Define v_i by $v_i = \varepsilon_i^{-1} u_i$, and observe that the v_i satisfy the equation

$$Lv_i + \varepsilon_i^{p_i - 1} K_i v_i^{p_i} = 0 . \tag{3.8}$$

Since $\{u_i\}$ is uniformly bounded on compact subsets of $M - \{P_1, \ldots, P_k\}$, we have from (3.4) a Harnack inequality for u_i on compact subsets of $M - \{P_1, \ldots, P_k\}$. Thus the v_i satisfy a Harnack inequality, and $v_i(Q) = 1$. Therefore the v_i are locally uniformly bounded on $M - \{P_1, \ldots, P_k\}$. From (3.8) we then get bounds on all derivatives of v_i away from $\{P_1, \ldots, P_k\}$. Therefore a subsequence, again denoted v_i, converges in C^3 norm on compact subsets of $M - \{P_1, \ldots, P_k\}$ to a smooth positive solution G of $LG = 0$ on $M - \{P_1, \ldots, P_k\}$. Since we are assuming $R(g_0) > 0$, G must be singular at one or more of the points $P_1, \ldots, P_k$. Suppose G is singular at $P_1, \ldots, P_\ell$. It then follows that G is a positive linear combination of (positive) fundamental solutions G_α with poles at P_α for $\alpha = 1, \ldots, \ell$. That is, there exist positive constants $a_1, \ldots, a_\ell$ such that $G = \sum_{\alpha=1}^{\ell} a_\alpha G_\alpha$. Let $x^1, \ldots, x^n$ be conformally flat coordinates centered at P_1. Let $\sigma > 0$ be a number which will be chosen small, and apply (3.2) with $u = u_i$ on D_σ. For a solution u of (3.4), we have $R(u^{4/(n-2)} g_0) = c(n)^{-1} K u^{-\delta}$ where $\delta = (n+2)/(n-2) - p$, and thus the left hand side of (3.2) can be written

$$c(n)^{-1} K \int_{D_\sigma} r \frac{\partial}{\partial r}(u^{-\delta})(\lambda_0 u)^{2n/(n-2)}$$

$$= -c(n)^{-1} K \delta (p+1)^{-1} \int r \frac{\partial}{\partial r}(u^{p+1}) \lambda_0^{2n/(n-2)} dx .$$

Since $r\frac{\partial}{\partial r} = \sum_i x^i \frac{\partial}{\partial x^i}$, we may integrate by parts to obtain

$$\int_{D_\sigma} r\frac{\partial}{\partial r}(u^{p+1})\lambda_0^{2n/(n-2)}dx = -\int_{D_\sigma} u^{p+1}\left(n + \frac{2n}{n-2} r \frac{\partial \log \lambda_0}{\partial r}\right)\lambda_0^{2n/(n-2)}dx$$
$$+ \sigma^n \int_{\partial D_\sigma} u^{p+1}\lambda_0^{2n/(n-2)}d\xi\,.$$

For σ small $n + \frac{2n}{n-2} r \frac{\partial \log \lambda_0}{\partial r} > 0$, and hence (3.2) implies the inequality

$$\frac{2n\sigma^n}{n-2}\int_{\partial D_\sigma}(\lambda_0 u)^2 T\left(\frac{\partial}{\partial r}, \frac{\partial}{\partial r}\right)d\xi \geq -c(n)^{-1}K\delta(p+1)^{-1}\sigma^n\int_{\partial D_\sigma} u^{p+1}\lambda_0^{2n/(n-2)}d\xi$$

for any solution u of (3.4). Applying this with $u = u_i$ and multiplying by ε_i^2 we get in the limit

$$\sigma^n \int_{\partial D_\sigma}(\lambda_0 G)^2 T\left(\frac{\partial}{\partial r}, \frac{\partial}{\partial r}\right)d\xi \geq 0 \tag{3.9}$$

where

$$T\left(\frac{\partial}{\partial r}, \frac{\partial}{\partial r}\right)$$

is given by (3.3) with $u = G$. Since the metric $G^{4/(n-2)}g_0 = (\lambda_0 G)^{4/(n-2)}\sum_i (dx^i)^2$ has zero scalar curvature, $\lambda_0 G$ is a positive Euclidean harmonic function on $D_\sigma - \{0\}$ which is singular at $x = 0$. It follows that $(\lambda_0 G)(x)$ has the expansion

$$(\lambda_0 G)(x) = a_1|x|^{2-n} + A + \alpha(x)$$

where $\alpha(x)$ is a harmonic function with $\alpha(0) = 0$. Using this expression in (3.9) we get $-A + O(\sigma) \geq 0$ by elementary calculation using (3.3). Since σ is arbitrarily small we get $A \leq 0$. On the other hand we have $G \geq a_1 G_1$, and

$$\lambda_0 G_1(x) = |x|^{2-n} + E(P_1) + O(|x|)$$

where $E(\cdot)$ is the energy function discussed in §4. Thus $A \geq a_1 E(P_1)$ which is strictly positive since (M, g_0) is not conformally equivalent to S^n. We discuss this positive energy statement in the next section. This contradiction shows that $u > 0$ on M.

The second step in our proof deals with the remaining case $u > 0$. In this case our argument is local. The sequence $\{u_i\}$ must be unbounded near one of the points $\{P_1, \ldots, P_k\}$, for otherwise we have uniform convergence. Assume that $\lim\{\sup_{B_\sigma(P_1)} u_i\} = \infty$ for any $\sigma > 0$. Since $u > 0$, the metrics $g_i = u_i^{4/(n-2)}g_0$ have uniformly bounded curvature away from the points $P_1, \ldots, P_k$. Let $x^1, \ldots, x^n$ be conformally flat coordinates centered at P_1. Let $\lambda_0(x) > 0$ be such that $g_0 = \lambda_0^{4/(n-2)}\sum(dx^i)^2$, and assume λ_0 is bounded above and below (locally). The functions $w_i = \lambda_0 u_i$ then satisfy

$$\Delta w_i + c(n)R_i w_i^{(n+2)/(n-2)} = 0$$

where Δ is the Euclidean Laplace operator and $R_i = R(g_i)$. In particular, w_i is superharmonic and by assumption w_i is bounded below on ∂D_σ for i large. Therefore w_i has a lower bound on D_σ. If the Ricci curvature of g_i were bounded in D_σ, then we can use the gradient estimate [6] on the solution w_i^{-1} of the equation $L_{g_i}(w_i^{-1}) = 0$. Note that this equation holds because $w_i^{-4/(n-2)} g_i$ is the Euclidean metric. The gradient estimate can be applied because of Proposition 3.2 which gives us a bound on R_i and $|\nabla_{g_0} R_i|$. We have

$$|\nabla_{g_i} R_i| = u_i^{-2/(n-2)} |\nabla_{g_0} R_i|$$

which is bounded since u_i has a lower bound. The gradient estimate then gives

$$|\nabla_{g_i} w_i^{-1}| \le c w_i^{-1} .$$

Writing this in terms of the Euclidean metric we have

$$|\partial(w_i^{-2/(n-2)})| \le c$$

where ∂ denotes the Euclidean gradient. Note that the gradient bound depends on the geodesic distance to ∂D_σ. Since w_i is bounded below we have $\sup_{D_{\sigma/2}} |\partial(w_i^{-2/(n-2)})| \le c\sigma^{-1}$. This implies an upper bound on w_i near 0 in terms of an upper bound at a fixed small distance from 0. Since w_i are converging away from 0, we get an upper bound independent of i. This contradiction shows that $\lim\{\sup_{D_\sigma} \|\mathrm{Ric}\,(g_i)\|_{g_i}\} = \infty$ for any $\sigma > 0$. Therefore we can choose a sequence of points $Q_i \to P_1$ such that

$$c_i = \sup_{D_\sigma} \|\mathrm{Ric}\,(g_i)\|_{g_i} = \|\mathrm{Ric}\,(g_i)\|_{g_i}(Q_i)$$

with $c_i \to \infty$. We then let $\overline{g}_i = c_i g_i$, and observe that we have

$$1 = \sup_{D_\sigma} \|\mathrm{Ric}\,(\overline{g}_i)\|_{\overline{g}_i} = \|\mathrm{Ric}\,(\overline{g}_i)\|_{\overline{g}_i}(Q_i) .$$

Thus we have $\overline{g}_i = \overline{w}_i^{4/(n-2)} \sum_j (dx^j)^2$ where $\overline{w}_i = c_i^{(n-2)/4} w_i$. Let x_i denote the point in D_σ corresponding to Q_i, so that we have $\lim x_i = 0$. Let

$$v_i(y) = \lambda_i^{(n-2)/2} w_i(\lambda_i y + x_i)$$
$$\overline{v}_i(y) = c_i^{(n-2)/4} v_i(y)$$

where we choose $\lambda_i = (\overline{w}_i(x_i))^{-2/(n-2)}$ so that $\overline{v}_i(0) = 1$. Notice that $\lambda_i \to 0$ so that v_i is defined on increasingly large balls in $\mathbf{R}^n$. Moreover, $v_i^{4/(n-2)} \sum (dy^j)^2$ is the pullback of g_i under the map $y \mapsto \lambda_i y + x_i$, and hence the scalar curvature and volume of $v_i^{4/(n-2)} \sum (dy^j)^2$ are bounded. Thus by Proposition 3.1, a subsequence of $\{v_i\}$ converges uniformly away from a finite set of points $y_1, \ldots, y_r \in \mathbf{R}^n$. In particular, v_i satisfies a Harnack inequality away from $y_1, \ldots, y_r$. Since the Ricci curvature of $\overline{v}_i^{4/(n-2)} \sum (dy^j)^2$ is bounded and the metric is complete, the Harnack inequality of [6] holds on unit geodesic balls. In particular, $\overline{v}_i$ remains bounded in a uniform neighborhood of $y = 0$, and 0 is distinct from $y_1, \ldots, y_r$. Therefore a subsequence of $\{\overline{v}_i\}$, again denoted $\{\overline{v}_i\}$, converges to a limit h. From the construction h is a positive harmonic function on $\mathbf{R}^n - \{y_1, \ldots, y_r\}$ with $h(0) = 1$. Moreover, the metric $h^{4/(n-2)} \sum (dy^j)^2$ has Ricci curvature of length one at $y = 0$ and in particular is not flat. It follows that h has at least two singularities in $S^n = \mathbf{R}^n \cup \{\infty\}$. Let $y_1, \ldots, y_s$ denote the singular points of h in S^n. It follows that h is a positive linear combination of fundamental solutions with poles at $y_1, \ldots, y_s$. Thus there are positive numbers $a_1, \ldots, a_s$ such that $h(y) = \sum_{\alpha=1}^{s} a_\alpha G_\alpha$ where $G_\alpha(y) = |y - y_\alpha|^{2-n}$ if $y_\alpha \in \mathbf{R}^n$ and $G_\alpha(y) \equiv 1$ if $y_\alpha = \infty$. Assume $y_1 \in \mathbf{R}^n$ so that

$$h(y) = a_1 |y - y_1|^{2-n} + A + \alpha(y)$$

where $\alpha(y_1) = 0$ and $A > 0$ because $s \geq 2$. Now the same argument as in the previous step, using (3.2), gives us a contradiction. This shows that our initial assumption of nonconvergence of $\{u_i\}$ is violated and we have completed the proof of Theorem 3.3.

There is an obvious question which is left unresolved by Theorem 3.3, and this is the question of whether one can remove the assumed bound on the energy $E(u)$ which is required in Theorem 3.3. It seems likely that the energy of solutions of (1.9) will be bounded by a constant depending only on g_0. Inequality (1.13) gives a very strong a priori integral estimate on solutions of (1.2) for $n = 3$. It may be possible to use this in place of the energy bound in Theorem 3.3.

4 The relevant positive energy theorems

In this section we give a discussion of the the total energy of an asymptotically flat n-manifold and discuss the positive energy theorems which are relevant to the Yamabe problem. Let (M^n, g) be a Riemannian manifold. (M, g) is said to be *asymptotically flat* if there is a compact subset $K \subset M$ such that $M - K$ is diffeomorphic to $\mathbf{R}^n - \{|x| \leq 1\}$, and a diffeomorphism $\Phi : M - K \to \mathbf{R}^n - \{|x| \leq 1\}$ such that, in the coordinate chart defined by Φ, we have $g = \sum_{i,j} g_{ij}(x)dx^i dx^j$ where $g_{ij}(x) = \delta_{ij} + O(|x|^{-p})$ as $x \to \infty$ for some $p > 0$. We also assume that

$$|x|\,|g_{ij,k}(x)| + |x|^2|g_{ij,k\ell}(x)| = O\left(|x|^{-p}\right)$$

where we use commas to denote partial derivatives as in §1. Assuming that $|R(g)| = O(|x|^{-q})$, $q > n$, and $p > (n-2)/2$ it is possible to define the total energy of M. To do this we recall the expression for $R(g)$ in the x-coordinates

$$R(g) = \sum_{i,j,k} g^{ij}\left(\Gamma^k_{ij,k} - \Gamma^k_{ik,j} + \sum_\ell (\Gamma^k_{k\ell}\Gamma^\ell_{ij} - \Gamma^k_{j\ell}\Gamma^\ell_{ik})\right)$$

$$\Gamma^k_{ij} = \frac{1}{2}\sum_m g^{km}(g_{im,j} + g_{jm,i} - g_{ij,m}).$$

Using the asymptotic assumptions we find

$$R(g) = \sum_{i,j}(g_{ij,ij} - g_{ii,jj}) + O\left(|x|^{-2p-2}\right).$$

Since $2p + 2 > n$ we therefore have the divergence term absolutely integrable near infinity. Thus the divergence theorem implies the existence of the following limit

$$\lim_{\sigma\to\infty}\int_{\{|x|=\sigma\}}\sum_{i,j}(g_{ij,i}\nu_j - g_{ii,j}\nu_j)\,d\xi(\sigma)$$

where $\nu = \sigma^{-1}x$ is the Euclidean unit normal to $\{|x| = \sigma\}$ and $d\xi(\sigma)$ denotes the Euclidean area element on $\{|x| = \sigma\}$. Moreover, the family of spheres $S_\sigma = \{|x| = \sigma\}$ may be replaced by any sequence of boundaries which go uniformly to infinity, and the limit will exist and have the same value (see [2]). We define the total energy $E = E(M, g)$ by

$$E = (4(n-1)w_{n-1})^{-1}\lim_{\sigma\to\infty}\int_{S_\sigma}\sum_{i,j}(g_{ij,i}\nu_j - g_{ii,j}\nu_j)d\xi(\sigma)$$

where $w_{n-1} = \mathrm{Vol}\,(S^{n-1}(1))$. The basic content of the positive energy theorem, or this special case of it, is that if $R(g) \geq 0$ on all of M, then $E \geq 0$. Moreover, $E = 0$ only if (M, g) is isometric to Euclidean space.

For a compact manifold (M, g) with $R(g) > 0$ we can make the following construction. Given a point $P \in M$, there is a positive fundamental solution G for the conformal Laplacian L with pole at P. If we normalize G so that

$$\lim_{Q\to P} d(P, Q)^{n-2}\, G(Q) = 1$$

where $d(\cdot,\cdot)$ is the Riemannian distance function for g, then G is unique. The manifold $(M-\{P\}, G^{4/(n-2)}g)$ is then asymptotically flat. If we let $y^1,\ldots,y^n$ denote a normal coordinate system for g centered at P, then we have $g_{ij} = \delta_{ij} + O(|y|^{-2})$. It is not difficult to show that $G(y) = |y|^{2-n} + O(|y|^{p+2-n})$ where p is any number less than two. If we let $x = |y|^{-2}y$, then we have the metric components of g in the x-coordinates given by $|x|^{-4} g_{ij}(|x|^{-2}x)$. In particular, if we let

$$G^{4/(n-2)}g = \sum_{i,j} \bar{g}_{ij}(x)dx^i dx^j$$

we have $\bar{g}_{ij}(x) = \delta_{ij} + O(|x|^{-p})$ as $x \to \infty$. Also we have $R(G^{4/(n-2)}g) = 0$ since $LG = 0$ on $M - \{P\}$. In particular, if $p > (n-2)/2$, then the total energy can be defined. Since $(n-2)/2 < 2$ for $n = 3, 4, 5$ we see that in these dimensions we can assign to each point $P \in M$ a number $E(P)$ which is the total energy of $(M - \{P\}, G^{4/(n-2)}g)$.

We now let (M^n, g) denote a general asymptotically flat manifold. We are going to present the minimal hypersurface proof of the positive energy theorem which is joint with S.T. Yau and appears in [25], [26]. Our presentation will simplify the original proofs in a few technical respects. It is convenient to first simplify the asymptotic behavior of g so that g is conformally flat near infinity. We carried out this argument for $n = 3$ in [27], and we present here the n-dimensional version.

Proposition 4.1. *Let (M, g) be asymptotically flat with $p > (n-2)/2$ and $q > n$. Assume also that $R(g) \geq 0$. For any $\varepsilon > 0$ there is a metric $\bar{g}$ such that $(M, \bar{g})$ is asymptotically flat and conformally flat near infinity with $R(\bar{g}) \equiv 0$ and such that $E(\bar{g}) \leq E(g) + \varepsilon$.*

Proof: We first observe that we may take $R(g) \equiv 0$, since generally, we can solve $Lu = 0$, $u > 0$ with $u \sim 1$ at infinity. In fact, we have $u(x) = 1 + A|x|^{2-n} + O(|x|^{1-n})$ where $A \leq 0$ since $0 < u < 1$ on M. (See [2] for the existence and expansion.) The metric $u^{4/(n-2)}g$ is then scalar flat and has total energy given by $E(g) + A \leq E(g)$. Thus g may be replaced by $u^{4/(n-2)}g$.

Now assume $R(g) \equiv 0$, and deform g near infinity to the Euclidean metric. To accomplish this, choose a function $\Psi_\sigma(x)$ with the properties, $\Psi_\sigma(x) = 1$ for $|x| \leq \sigma$, $\Psi_\sigma(x) = 0$ for $|x| \geq 2\sigma$, Ψ_σ is a decreasing function of $|x|$, and $\sigma|\Psi'_\sigma| + \sigma^2|\Psi''_\sigma| \leq c$. Now consider the metric ${}^{(\sigma)}g$ given by ${}^{(\sigma)}g = \psi_\sigma g + (1 - \Psi_\sigma)\delta$ where $\delta = \sum_{i,j}\delta_{ij}dx^i dx^j$ denotes the Euclidean metric. Observe that ${}^{(\sigma)}g = \delta + O(|x|^{-p})$ uniformly in σ for σ large, and also $R({}^{(\sigma)}g) = O(|x|^{-2-p})$ for $\sigma \leq |x| \leq 2\sigma$ uniformly in σ. In particular we have

$$\int_M |R({}^{(\sigma)}g)|^{n/2} d\omega_g = O(\sigma^{-np/2}),$$

and so for σ large there is a unique solution u_σ of $L_\sigma u_\sigma = 0$, $u_\sigma > 0$, $u_\sigma \to 1$ as $|x| \to \infty$. (See [2] for the existence.) The metric ${}^{(\sigma)}\bar{g} = u_\sigma^{4/(n-2)}\,{}^{(\sigma)}g$ is then scalar flat and conformally Euclidean near infinity. We show $\lim_{\sigma\to\infty} E({}^{(\sigma)}\bar{g}) = E(g)$ and then for σ sufficiently large the metric ${}^{(\sigma)}\overline{g}$ will give the desired metric. From the uniform decay estimates on u_σ and ${}^{(\sigma)}g$, we see that given $\varepsilon > 0$ there is a σ_0 independent of σ such that

$$|E({}^{(\sigma)}\overline{g}) - (4(n-1)w_{n-1})^{-1} \int_{S_{\sigma_0}} \sum_{i,j} ({}^{(\sigma)}\overline{g}_{ij,i}\nu_j - {}^{(\sigma)}\overline{g}_{ii,j}\nu_j) d\xi(\sigma_0)| \leq \frac{\varepsilon}{3}$$

$$|E(g) - (4(n-1)w_{n-1})^{-1} \int_{S_{\sigma_0}} \sum_{i,j} (g_{ij,i}\nu_j - g_{ii,j}\nu_j) d\xi(\sigma_0)| \leq \frac{\varepsilon}{3}.$$

On the other hand, we have $\lim_{\sigma\to\infty} u_\sigma = 1$ on compact subsets of M, and hence the two surface integrals above are within $\varepsilon/3$ when σ is sufficiently large. Thus we get $|E({}^{(\sigma)}\overline{g}) - E(g)| < \varepsilon$ for σ large. This completes the proof of Proposition 4.1.

Note that if (M, g) is asymptotically flat and conformally flat near infinity we have $g_{ij} = h^{4/(n-2)}(x)\delta_{ij}$ for $|x|$ large where $h(x) \to 1$ as $x \to \infty$. If $R(g) \equiv 0$, then h is a harmonic function for $|x|$ large and hence $h(x) = 1 + E|x|^{2-n} + O(|x|^{1-n})$ where we have normalized the energy so that E is the energy of the metric $h^{4/(n-2)}\delta$. Thus by Proposition 4.1 we may assume g to be of this form.

Theorem 4.2. *Let (M, g) be asymptotically flat with $p > (n-2)/2$, $q > n$, and $R(g) \geq 0$ on M. Then $E(g) \geq 0$ and $E(g) = 0$ only if (M, g) is isometric to $(\mathbf{R}^n, \delta)$.*

We will give the proof of this theorem for $n \leq 7$. This proof can be extended to arbitrary dimensions with an additional technical complication arising from singular sets of area minimizing hypersurfaces which appear for $n \geq 8$. We do not deal with this here, but leave it to a forthcoming work of the author and S.T. Yau. In any case, this is not required for the Yamabe problem as the remaining case of locally conformally flat manifolds of arbitrary dimension has been treated by a different argument in [29]. For the case in which M is a spin manifold a different proof of Theorem 4.2 was given by E. Witten [34]. This proof was carried over to arbitrary dimensions in [15].

Proof of Theorem 4.2: We first show that $E \geq 0$. Suppose on the contrary that $E < 0$. Then by Proposition 4.1 we may assume $R(g) \equiv 0$ and $g_{ij} = h^{4/(n-2)}\delta_{ij}$ where $h(x)$ has the expansion $h(x) = 1 + E|x|^{2-n} + (|x|^{1-n})$ for x large. It will be convenient to have $R(g) > 0$ on M. This can be accomplished by replacing g by $u^{4/(n-2)}g$ where u satisfies $Lu = -g$ with $g > 0$ on M, g small and g decaying rapidly. The solution u will then satisfy $u(x) = 1 + \delta|x|^{2-n} + O(|x|^{1-n})$ with δ arbitrarily small. Thus the negativity of the energy is preserved. We compute the divergence of the unit vector field $\eta = h^{-2/(n-2)}\frac{\partial}{\partial x^n}$ with respect to g. We find

$$\begin{aligned} \operatorname{div}_g(\eta) &= h^{-2n/(n-2)} \frac{\partial}{\partial x^n}(h^{2n/(n-2)}h^{-2/(n-2)}) \\ &= \frac{2(n-1)}{n-2} E \frac{\partial}{\partial x^n}\left(|x|^{2-n}\right) + O\left(|x|^{-n}\right) \\ &= -2(n-1) E \frac{x^n}{|x|^n} + O\left(|x|^{-n}\right) \end{aligned}$$

In particular we see that $\operatorname{div}_g(\eta) > 0$ for $x^n \geq a_0$ and $\operatorname{div}_g(-\eta) > 0$ for $x^n \leq -a_0$ for some constant a_0. Now let σ be a large radius, and let $\Gamma_{\sigma,a}$ denote the $(n-2)$-dimensional sphere

$$\Gamma_{\sigma,a} = \{x = (x', x^n) : |x'| = \sigma, x^n = a\}.$$

Let C_σ denote the $(n-1)$ dimensional cylinder $C_\sigma = \{(x', x^n) : |x'| = \sigma\}$. We orient $\Gamma_{\sigma,a}$ as the boundary of the portion of C_σ lying below $\Gamma_{\sigma,a}$. Let $\Sigma_{\sigma,a}$ be an $(n-1)$-dimensional surface of least area with $\partial\Sigma_{\sigma,a} = \Gamma_{\sigma,a}$. The cylinder C_σ bounds an interior region Ω_σ in M, and $\Sigma_{\sigma,a} \subset \Omega_\sigma$. Since $n \leq 7$, $\Sigma_{\sigma,a}$ will be free of singularities (see [11,30] for relevant results on the Plateau problem). For any σ, let

$$V(\sigma) = \min\{\operatorname{Vol}(\Sigma_{\sigma,a}) : a \in [-a_0, a_0]\}$$

where we note that the function $a \mapsto \operatorname{Vol}(\Sigma_{\sigma,a})$ is continuous. We now assert that there exists $a = a(\sigma) \in (-a_0, a_0)$ such that $\operatorname{Vol}(\Sigma_{\sigma,a}) = V(\sigma)$. To show that $a(\sigma) < a_0$, write $\Sigma_{\sigma,a} = (\partial\Omega_{\sigma,a}) \cap \Omega_\sigma$ where $\Omega_{\sigma,a}$ is the subregion of Ω_σ lying below $\Sigma_{\sigma,a}$. Let

$$U_{\sigma,a} = \{(x', x^n) \in \Omega_{\sigma,a} : x^n > a_0 - \delta\}$$

where δ is chosen so small that $\operatorname{div}_g(\eta) > 0$ for $x^n > a_0 - \delta$. We show that $U_{\sigma,a} = \emptyset$ by applying the divergence theorem in $U_{\sigma,a}$. Since η is tangent to C_σ, we get

$$\int_{\Sigma_{\sigma,a}\cap\{x^n \geq a_0-\delta\}} \langle \eta, \nu\rangle_g \, d\mathcal{H}^{n-1} - \operatorname{Vol}(\Omega_{\sigma,a} \cap \{x^n = a_0 - \delta\}) > 0$$

provided $U_{\sigma,a} \neq \emptyset$. Here ν denotes the unit normal of $\Sigma_{\sigma,a}$. Thus we may apply the Schwarz inequality to assert

$$\operatorname{Vol}(\Omega_{\sigma,a} \cap \{x^n = a_0 - \delta\}) < \operatorname{Vol}(\Sigma_{\sigma,a} \cap \{x^n \geq a_0 - \delta\}).$$

Therefore, if $U_{\sigma,a} \neq \emptyset$, then the hypersurface Σ given by

$$\Sigma = \partial(\Omega_{\sigma,\delta} \cap \{x^n < a_0 - \delta\}) \cap \Omega_\sigma$$

has smaller volume than $\Sigma_{\sigma,a}$ and $\partial\Sigma = \Gamma_{\sigma,a_1}$ where $a_1 = \min\{a, a_0 - \delta\}$. This contradiction shows that $U_{\sigma,a} = \emptyset$ and in particular $a(\sigma) \leq a_0 - \delta$. An analogous argument shows that $a(\sigma) \geq -a_0 + \delta$ for some $\delta > 0$.

Let $\Sigma_\sigma = \Sigma_{\sigma,a(\sigma)}$ be one of the hypersurfaces which realizes the minimum volume $V(\sigma)$. Let $\mathbf{X}_1$ be a fixed vector field on M which is equal to $\frac{\partial}{\partial x^n}$ outside a compact set. Let $\mathbf{X}_0$ be a vector field of compact support, and let $\mathbf{X} = \mathbf{X}_0 + \alpha\mathbf{X}_1$ where $\alpha \in \mathbf{R}$. Let F_t be the one parameter group of diffeomorphisms generated by $\mathbf{X}$ (or alternatively any curve of diffeomorphisms whose tangent vector at $t = 0$ is $\mathbf{X}$). If σ is sufficiently large that the support of $\mathbf{X}_0$ is compactly contained in Ω_σ, then $\mathbf{X}$ gives a valid variation of Σ_σ; that is, we have

$$\frac{d}{dt}\operatorname{Vol}(F_t(\Sigma_\sigma))\Big|_{t=0} = 0, \quad \frac{d^2}{dt^2}\operatorname{Vol}(F_t(\Sigma_\sigma))\Big|_{t=0} \geq 0.$$

The second variation is the integral of the function $F_{\mathbf{X},\sigma}$ given by

$$F_{\mathbf{X},\sigma}(P) = \frac{d^2}{dt^2} \|(F_t)_*(T_P\Sigma_\sigma)\|_g \Big|_{t=0}$$

where $T_P\Sigma_\sigma$ denotes the oriented tangent plane of Σ_σ at P, and $(F_t)_*$ denotes the differential of the map F_t. For $|x|$ large we have $F_{\mathbf{X},\sigma}(x) = O(|x|^{-n})$ uniformly in σ because of the decay property which is assumed on g. The regularity theory implies that outside a fixed compact set Σ_σ is the graph of a function $f_\sigma(x')$, $x' = (x^1, \ldots, x^{n-1})$ having bounded gradient. We choose a sequence $\sigma_i \to \infty$ such that $\{\Sigma_{\sigma_i}\}$ converges to a limiting area minimizing hypersurface $\Sigma \subset M$. Because of the uniform decay condition on $F_{\mathbf{X},\sigma_i}$, we get $\int_\Sigma F_{\mathbf{X}}\, d\mathcal{H}^{n-1} \geq 0$ where

$$F_{\mathbf{X}}(P) = \frac{d^2}{dt^2} \|(F_t)_*(T_P\Sigma)\| \Big|_{t=0}$$

and $\mathbf{X} = \mathbf{X}_0 + \alpha\mathbf{X}_1$ for vector fields $\mathbf{X}_0$ of compact support and $\mathbf{X}_1$ fixed as above. Outside a compact subset of M the surface Σ is represented as the graph of a function $f(x')$ of bounded gradient. In fact, we easily get $|\partial f|(x) = (|x'|^{-1})$ from the regularity theory since we have a uniform bound on f, $|f(x')| \leq a_0$. On the other hand f satisfies the minimal surface equation

$$\sum_{i,j} \left(\delta_{ij} - \frac{f_{,i} f_{,j}}{1 + |\partial f|^2}\right) f_{,ij} + \sqrt{1 + |\partial f|^2}\, \frac{\partial}{\partial \nu_0} \log h = 0$$

where

$$h(x) = 1 + E|x|^{2-n} + O\left(|x|^{1-n}\right),$$

and

$$\nu_0 = (1 + |\partial f|^2)^{-1/2}(-\partial f, 1)$$

is the Euclidean unit normal vector. Applying linear theory (see [10]) we get $f(x') = a + O(|x'|^{3-n})$ for $n \geq 4$, and $f(x') = a + O(|x'|^{-1})$ for $n = 3$ for some constant a. The function F_X can be calculated in terms of the geometry of Σ (see [30])

$$F_{\mathbf{X}} = -\sum_{i=1}^{n-1} \langle R(\mathbf{X}, e_i)\mathbf{X}, e_i\rangle + \operatorname{div}_M Z + (\operatorname{div}_M \mathbf{X})^2$$

$$+ \sum_{i=1}^{n-1} |(D_{e_i}\mathbf{X})^\perp|^2 - \sum_{i,j=1}^{n-1} \langle e_i, D_{e_j}\mathbf{X}\rangle \langle e_j, D_{e_i}\mathbf{X}\rangle$$

where $Z = D_{\frac{\partial}{\partial t}} \frac{\partial F_t}{\partial t}$ is the acceleration vector field of the deformation, and D is used to denote covariant differentiation in M with respect to g. We use the notation

$$\operatorname{div}_M \mathbf{X} = \sum_{i=1}^{n-1} \langle D_{e_i}\mathbf{X}, e_i\rangle$$

where $\mathbf{X}$ is a (not necessarily tangent) vector field along Σ and $e_1, \ldots, e_{n-1}$ denotes an orthonormal basis for the tangent space to Σ. We write $\mathbf{X} = \hat{\mathbf{X}} + \varphi\nu$ where $\hat{\mathbf{X}}$ is tangent to Σ

and ν is the unit normal. Similarly $Z = \hat{Z} + \psi\nu$. Since Σ is minimal we have $\operatorname{div}_M \chi\nu = 0$ for any function χ. We then have

$$F_{\mathbf{X}} = -\varphi^2 \mathrm{Ric}\,(\nu,\nu) - \varphi^2\|B\|^2 + \|\nabla\varphi\|^2 + G$$

where G is given by

$$\begin{aligned} G = & -2\sum_{i=1}^{n-1} \varphi\langle R(\hat{\mathbf{X}}, e_i)\nu, e_i\rangle - \sum_{i=1}^{n-1} \varphi\langle R(\hat{\mathbf{X}}, e_i)\hat{\mathbf{X}}, e_i\rangle \\ & + \operatorname{div}_M \hat{Z} + (\operatorname{div}_M \hat{\mathbf{X}})^2 - 2B(\nabla\varphi, \hat{\mathbf{X}}) + \sum_{i=1}^{n-1} B(e_i, \hat{\mathbf{X}})^2 \\ & - 2\varphi \sum_{i,j=1}^{n-1} b_{ij}\hat{\mathbf{X}}_{i;j} - \sum_{i,j=1}^{n-1} \hat{\mathbf{X}}_{i;j}\hat{\mathbf{X}}_{j;i}\,. \end{aligned}$$

In these formulas we work in an *orthonormal* frame, $B(\cdot,\cdot)$ denotes the second fundamental form given by $B(V,W) = \langle D_V W, \nu\rangle$ for tangent vector fields V, W. We let $b_{ij} = B(e_i, e_j)$ in our orthonormal basis, and for a tangent vector field $V = \sum V_i e_i$, $V_{i;j}$ denotes the covariant derivative in the induced metric on Σ. Any term which involves Z or $\hat{\mathbf{X}}$ must reduce to a boundary term. If $D \subset \Sigma$ is a bounded domain, we see

$$\begin{aligned} \int_D G\, d\mathcal{H}^{n-1} = \int_{\partial D} \Big\{ & (\operatorname{div}\hat{\mathbf{X}})\,\langle\hat{\mathbf{X}}, \eta\rangle - \sum_{i,j} \hat{\mathbf{X}}_{i;j}\hat{\mathbf{X}}_j\eta_i \\ & -2\varphi\sum_{i,j} b_{ij}\hat{\mathbf{X}}_i\eta_j + \langle\hat{Z}, \eta\rangle \Big\}\, d\mathcal{H}^{n-2} \end{aligned}$$

where η is the outward normal to ∂D in Σ. To see that the interior terms drop out one must use the Gauß and Codazzi equations as well as the Ricci formula. For $\sigma > 0$, let $D_\sigma = \Omega_\sigma \cap \Sigma$ where Ω_σ is the interior region bounded by C_σ as above. From the decay conditions on f and h one checks that each of the boundary terms above decays faster than σ^{2-n}, and hence the boundary term tends to zero as $\sigma \to \infty$. Therefore we conclude

$$\int_\Sigma (\mathrm{Ric}\,(\nu,\nu) + \|B\|^2)\varphi^2 d\mathcal{H}^{n-1} \le \int_\Sigma |\nabla\varphi|^2 d\mathcal{H}^{n-1} \tag{4.1}$$

where $\varphi = \langle \mathbf{X}, \nu\rangle$ and $\mathbf{X} = \mathbf{X}_0 + \alpha\mathbf{X}_1$ as above. Since $\mathbf{X}$ can be chosen to be arbitrary except that $\mathbf{X} = \alpha\frac{\partial}{\partial x^n}$ outside a compact set, we see that φ is arbitrary except that

$$\varphi = \alpha\left\langle \frac{\partial}{\partial x^n}, \nu\right\rangle = \alpha h^{2/(n-2)}(1 + |\partial f|^2)^{-1/2}$$

outside a compact set for a constant α. Since $\varphi - \alpha = O\left(|x'|^{2-n}\right)$ we see that $\varphi - \alpha$ has finite energy and therefore we can take φ to be any function for which $\varphi - \alpha$ has compact support (or finite energy) for some constant α. As in [28] we can use the Gauß equation to write

$$\mathrm{Ric}\,(\nu,\nu) + \|B\|^2 = \frac{1}{2}R_M - \frac{1}{2}R_\Sigma + \frac{1}{2}\|B\|^2 \tag{4.2}$$

where R_Σ is the (intrinsic) scalar curvature of Σ in the induced metric.

To complete the proof, we first suppose $n = 3$ and choose $\varphi \equiv 1$ in (4.1) to obtain $\frac{1}{2}\int_\Sigma R_\Sigma \, d\mathcal{H}^2 > 0$. Now $\frac{1}{2} R_\Sigma$ is simply the Gaussian curvature of Σ. The decay estimates for f, h easily imply that the total geodesic curvature of ∂D_σ converges to 2π where $D_\sigma = \Sigma \cap \Omega_\sigma$. Therefore we may apply the Gauß–Bonnet theorem on D_σ and let σ tend to infinity to get

$$\frac{1}{2}\int_\Sigma R_\Sigma \, d\mathcal{H}^2 = 2\pi\chi(\Sigma) - 2\pi .$$

Since $\chi(\Sigma) \leq 1$ for an open surface Σ, the right hand side is nonpositive. This contradicts the previous inequality and completes the proof for $n = 3$.

Now suppose $n \geq 4$, and observe that the induced metric $\bar{g}$ on Σ satisfies (in terms of coordinates $x^1, \ldots, x^{n-1}$)

$$\bar{g}_{ij} = h(x', f(x'))^{4/(n-2)}(\delta_{ij} + f_{,i} f_{,j}) = \delta_{ij} + O\left(|x|^{2-n}\right) .$$

Therefore $(\Sigma, \bar{g})$ is asymptotically flat and has energy zero. Inequality (4.1) together with (4.2) and the inequality $R_M \geq 0$ imply that the lowest Dirichlet eigenvalue for $L_{\bar{g}}$ on any compact domain in Σ is positive because $c(n) = \frac{n-2}{4(n-1)} < \frac{1}{2}$ for $n \geq 3$. Linear theory then enables us to solve $L_{\bar{g}} u = 0$ on Σ, $u > 0$ on Σ, and $u \to 1$ at infinity. Moreover, u has the expansion

$$u(x') = 1 + E_0 |x'|^{3-n} + O\left(|x'|^{2-n}\right) .$$

In particular, $u - 1$ has finite energy on Σ, and we may take $\varphi = u$ in (4.1). Using (4.2) and the fact that $R_M > 0$ we get

$$-\int_\Sigma R_\Sigma u^2 \, d\mathcal{H}^{n-1} < 2\int_\Sigma |\nabla u|^2 \, d\mathcal{H}^{n-1} \leq c(n)^{-1} \int_\Sigma |\nabla u|^2 \, d\mathcal{H}^{n-1} .$$

We may then write

$$\begin{aligned}\int_\Sigma |\nabla u|^2 \, d\mathcal{H}^n &= \lim_{\sigma\to\infty} \int_{D_\sigma} |\nabla u|^2 \, d\mathcal{H}^n \\ &= -c(n) \lim_{\sigma\to\infty} \int_{D_\sigma} R_\Sigma u^2 \, d\mathcal{H}^{n-1} + \lim_{\sigma\to\infty} \int_{\partial D_\sigma} u \frac{\partial u}{\partial \eta} \, d\mathcal{H}^{n-1}\end{aligned}$$

where η denotes the outward unit normal to ∂D_σ. From the expansion for u we then find $E_0 < 0$. Thus $(\Sigma, u^{4/(n-3)}\bar{g})$ is asymptotically flat, has zero scalar curvature, and negative total energy. The contradiction now follows inductively from $n = 3$. This completes the proof that $E \geq 0$. The statement that $E = 0$ only if (M, g) is isometric to $\mathbf{R}^n$ is proven in [23, Lemma 3 and Proposition 2]. We omit the details. This completes the proof of Theorem 4.2.

5 Noncompact manifolds and weak solutions

One of the results of [29] is that a simply connected, complete, locally conformally flat manifold (M, g) with $R(g) \geq 0$ is conformally diffeomorphic to a domain $\Omega \subset S^n$ with the Hausdorff dimension of $S^n - \Omega$ being at most $(n-2)/2$. In particular, any compact locally conformally flat manifold (M, g_0) with $R(g_0) \geq 0$ is conformally covered by a simply connected domain $\Omega \subset S^n$ with $\dim(S^n - \Omega) \leq (n-2)/2$. Thus by lifting solutions of (1.9) from M to Ω we get solutions $u > 0$ on Ω of the equation

$$Lu + \frac{n(n-2)}{4} u^{(n+2)/(n-2)} = 0 \tag{5.1}$$

where $Lu = \Delta_{S^n} u - \frac{n(n-2)}{4} u$. These solutions satisfy the "boundary condition" that $(\Omega, u^{4/(n-2)} g_0)$ is a complete Riemannian manifold. Here we take g_0 to be the metric on the unit sphere. The theorem of Obata discussed in §1 classifies the global regular solutions of (5.1). The first example of a domain Ω arising from the above construction is $S^n - \{P, Q\}$ for two points $P, Q \in S^n$. After a conformal transformation, we can take $Q = -P$ and think of $S^n = \mathbf{R}^n \cup \{\infty\}$ with $P = 0$, $Q = \infty$. We explicitly analyzed the solutions of (5.1) for this domain Ω in §2. In general, any domain Ω arising as the universal cover (or any covering) of a compact manifold is invariant under a discrete subgroup Γ of the conformal group of S^n and is the domain of discontinuity of this group. From Kleinian group theory we know that if the limit set $\Lambda = S^n - \Omega$ contains more than two points, then it must contain a Cantor set. It is a theorem in [29] that for a domain Ω which covers a compact manifold, the quotient manifold Ω/Γ has a conformal metric of positive scalar curvature if and only if the Hausdorff dimension of $S^n - \Omega$ is less than $(n-2)/2$.

Generally, if u is a solution of (5.1) on a domain $\Omega \subset S^n$ such that $(\Omega, u^{4/(n-2)} g_0)$ is a complete manifold, then it is shown in [29] that u is integrable on S^n to the power $(n+2)/(n-2)$ and that u defines a global weak solution of (5.1) on S^n. Thus the problem of constructing complete solutions of (5.1) on Ω is closely related to the problem of constructing weak solutions of (5.1) on S^n with prescribed singular set $\Lambda = S^n - \Omega$. We have seen that many solutions of (5.1) exist which are singular at two specified points; in fact, such solutions can be classified. The question of specifying more than two singular points has been posed in various contexts over the years. (Solutions do not exist with one singular point.) An obvious approach to this problem would be to fix the asymptotic behavior near k specified points of S^n and to construct a solution which is essentially a compact perturbation of a given function with the correct asymptotics. The difficulties in this approach are apparent from analysis of the solutions singular at $0, \infty$. Let $x \in \mathbf{R}^n$, $t = \log|x|$ as in §2. The simplest solution of (2.2) is the constant solution $u(t) \equiv u_0$. This gives rise to the solution $v(x) = u_0|x|^{-(n-2)/2}$ in $\mathbf{R}^n - \{0\}$ of the equation $\Delta v + \frac{n(n-2)}{4} v^{(n+2)/(n-2)} = 0$ which is equivalent to equation (5.1). If we consider solutions which are near u_0 on a large piece of $\mathbf{R} \times S^{n-1}$, then we would expect the linearized equation at u_0 to dictate their behavior. The linearized operator is $\mathcal{L}\eta = \Delta\eta + (n-2)\eta$ where Δ is with respect to the metric $dt^2 + d\xi^2$ on the cylinder. In particular, we see that zero is embedded in the continuous spectrum for $\mathcal{L}$ on $\mathbf{R} \times S^{n-1}$. Thus controlling $\mathcal{L}^{-1}$ on large regions of $\mathbf{R} \times S^{n-1}$ will be a difficult problem. It is not known whether solutions exist with asymptotic behavior given by the constant solution u_0. In [24] we proved a general existence theorem for weak solutions which implies that one can specify any k points of S^n and construct solutions singular at these points and asymptotic to solutions described in §2 with α near one. Roughly speaking, the spectrum of the linearized operator for such solutions ($\alpha \approx 1$) contains a small interval near 0, and the spectral subspace corresponding to this interval imposes an infinite number of geometric "balancing" conditions on the way in which spherical pieces of solutions are attached. We refer the reader to [24] for details.

References

[1] T. Aubin, *The scalar curvature, Differential Geometry and Relativity*, (Cahen and Flato, eds.), Reidel, Dordrecht 1976.

[2] R. Bartnik, *The mass of an asymptotically flat manifold*, Comm. Pure Appl. Math. **39** (1986), 661-693.

[3] M. Berger and D. Ebin, *Some decompositions of the space of symmetric tensors on a Riemannian manifold*, J. Diff. Geom. **3** (1969), 379–392.

[4] A. Besse, *Einstein Manifolds*, Springer–Verlag, Berlin, Heidelberg 1987.

[5] L. Caffarelli, B. Gidas and J. Spruck, *Asymptotic symmetry and local behavior of semilinear elliptic equations with critical Sobolev growth*, Preprint 1988.

[6] S.Y. Cheng and S.T. Yau, *Differential equations on Riemmanian manifolds and their geometric application*, Comm. Pure Appl. Math. **28** (1975), 333–354.

[7] H.I. Choi and R. Schoen, *The space of minimal embeddings of a surface into a three-dimensional manifold of positive Ricci curvature*, Invent. Math. **81** (1985), 387–394.

[8] A. Fischer and J. Marsden, *The manifold of conformally equivalent metrics*, Can. J. Math. **29** (1977), 193–209.

[9] B. Gidas, W.M. Ni and L. Nirenberg, *Symmetry and related properties via the maximum principle*, Comm. Math. Phys. **68** (1979), 209–243.

[10] D. Gilbarg and N. Trudinger, *Elliptic Partial Differential Equations of Second Order*, Springer–Verlag, Berlin, Heidelberg, New York, Tokyo 1983.

[11] E. Giusti, *Minimal Surfaces and Functions of Bounded Variation*, Birkhäuser, Boston, Basel, Stuttgart 1984.

[12] M. Gromov, *Pseudo-holomorphic curves in symplectic manifolds*, Invent. Math. **82** (1985), 307–347.

[13] M. Gromov and H.B. Lawson, *Positive scalar curvature and the Dirac operator*, IHES Publ. Math. **58** (1983), 83–196.

[14] N. Koiso, *On the second derivative of the total scalar curvature*, Osaka J. Math. **16** (1979), 413-421.

[15] J. Lee and T. Parker, *the Yamabe problem*, Bull. Amer. Math. Soc. **17** (1987), 37–81.

[16] J. Lelong-Ferrand, *Transformations conformes et quasiconformes des variétés riemanniennes; applications à la démonstration d'une conjecture de A. Lichnerowicz*, C.R. Acad. Sci. Paris **269** (1969), 583-586.

[17] P.L. Lions, *The concentration-compactness principle in the calculus of variations*, Ann. Inst. Henri Poincaré Analyse Nonlinéaire **1** (1984), 223-284.

[18] M. Obata, *The conjectures on conformal transformations of Riemannian manifolds*, J. Diff. Geom. **6** (1972), 247-258.

[19] T. Parker and C. Taubes, *On Witten's proof of the positive energy theorem*, Comm. Math. Phys. **84** (1982), 223-238.

[20] S. Pohozaev, *Eigenfunctions of the equation $\Delta u + \lambda f(u) = 0$*, Soviet Math. Dokl. **6** (1965), 1408-1411.

[21] E. Rodemich, *The Sobolev inequalities with best possible constants*, Analysis Seminar at the California Institute of Technology, 1966.

[22] J. Sacks and K. Uhlenbeck, *The existence of minimal immersions of 2-spheres*, Ann. Math. **113** (1981), 1-24.

[23] R. Schoen, *Conformal deformation of a Riemmanian metric to constant scalar curvature*, J. Diff. Geom. **20** (1984), 479-495.

[24] R. Schoen, *The existence of weak solutions with prescribed singular behavior for a conformally invariant scalar equation*, Comm. Pure Appl. Math. **41** (1988), 317-392.

[25] R. Schoen and S.T. Yau, *On the proof of the positive mass conjecture in General Relativity*, Comm. Math. Phys. **65** (1979), 45-76.

[26] R. Schoen and S.T. Yau, *Proof of the positive action conjecture in quantum relativity*, Phys. Rev. Let. **42** (1979), 547-548.

[27] R. Schoen and S.T. Yau, *Positivity of the total mass of general space-time*, Phys. Rev. Let. **43** (1979), 1457-1459.

[28] R. Schoen and S.T. Yau, *On the structure of manifolds with positive scalar curvature*, Manuscripta Math. **28** (1979), 159-183.

[29] R. Schoen and S.T. Yau, *Conformally flat manifolds, Kleinian groups, and scalar curvature*, Invent. Math. **92** (1988), 47-71.

[30] L. Simon, *Lectures on Geometric Measure Theory*, Centre for Mathematical Analysis, Australian National University, 1984.

[31] G. Talenti, *Best constant in Sobolev inequality*, Ann. Mat. Pura. Appl. **110** (1976), 353-372.

[32] N. Trudinger, *Remarks concerning the conformal deformation of Riemannian structures on compact manifolds*, Ann. Scuola Norm. Sup. Pisa Cl. Sci. (3) **22** (1968), 265-274.

[33] K. Uhlenbeck, *Connections with L^p-bounds on curvature*, Comm. Math. Phys. **83** (1982), 31-42.

[34] E. Witten, *A new proof of the positive energy theorem*, Comm. Math. Phys. **80** (1981), 381-402.

[35] H. Yamabe, *On a deformation of Riemannian structures on compact manifolds*, Osaka J. Math. **12** (1960), 21-37.

A Classical Variational Approach to Teichmüller Theory

A.J. Tromba

§ I Introduction

The purpose of these two lectures is to give a proof of Teichmüller's famous result on the structure of the moduli space for compact Riemann surfaces that now bears his name.

The classical approach employed by Teichmüller, Ahlfors, Bers and their students is fundamentally linked to the notion of quasi-conformal and extremal quasi-conformal mappings. These objects are minima of a variational problem involving supremum norms and consequently the classical theory sits on the outside of many other mathematical developments in recent decades, particularly in the variational calculus and in Riemannian geometry.

Teichmüller's space of moduli has come to play an important role not only as a subject in its own right but as an important tool in the study of minimal and H-surfaces (surfaces constant mean curvature) of higher genus, as well as a fundamental tool in the study of string theory in physics. In the present lectures we shall outline an approach to the moduli question that employs only mainstream ideas in the above two fields.

Most of the ideas in these lectures were developed jointly with Arthur Fischer [5], [6], [7], [8], [9]. The author wishes to thank CIME and especially the organizer of this variational meeting, Mariano Giaquinta for a truly delightful and stimulating meeting.

§ 1 What is Teichmüller's Space?

For the rest of these lectures M will denote a C^∞ smooth compact oriented surface without boundary of genus greater than one. It is well known that such an M admits a complex structure. A complex structure c is a collection of coordinate mappings $\{\varphi_i, U_i\}$, $\bigcup_i U_i = M$, U_i open, $\varphi_i : U_i \longrightarrow R^2$ such that when defined $\varphi_i \circ \varphi_j^{-1}$ is holomorphic.

Let $\mathcal{C}$ denote the space of all such complex structures. $\mathcal{C}$ is necessarily an infinite set. Now suppose $f : M \longrightarrow M$ is a C^∞ self mapping of M with a C^∞ inverse f^{-1} (such an f is a diffeomorphism). Using f we can construct a new complex structure f^*c, with $f^*c = \{\varphi_i \circ f, f^{-1}(U_i)\}$. It follows easily that f^*c is a complex structure iff c is a complex structure. Moreover if (M, f^*c) and (M, c) denote M with the complex structures f^*c and c respectively,

$$f : (M, f^*c) \longrightarrow (M, c)$$

is a holomorphic mapping. Thus one would like to identify these two complex structures. Another way of saying this is the following: Let $\mathcal{D}$ be the set of all such diffeomorphisms f. Then $\mathcal{D}$ acts on $\mathcal{C}$ by sending $c \longrightarrow f^*c$. Riemann's original problem was to study the quotient set $\mathcal{C}/\mathcal{D}$. Riemann conjectured that $\mathcal{C}/\mathcal{D} = \mathcal{R}(M)$ is a space of dimension $6(\text{genus } M) - 6$. $\mathcal{R}(M)$ is now called the Riemann space of moduli and its structure is still not completely understood.

Teichmüller's program was to tackle the study of this space by a twofold process. First he want to study the space $\mathcal{C}/\mathcal{D}_0$, where $\mathcal{D}_0$ are those diffeomorphisms homotopic to the identity. The space $\mathcal{T}(M) = \mathcal{C}/\mathcal{D}_0$ is now called Teichmüller's moduli space. The quotient group $\Gamma = \mathcal{D}/\mathcal{D}_0$ (the surface moduli group) is known to be an infinite discrete group.

It follows that Riemann's moduli space $\mathcal{R}(M) = \mathcal{T}(M)/\Gamma$,

is the quotient space of Teichmüller's moduli space and the surface moduli group. Teichmüller's remarkable result is that $\mathcal{T}(M)$ has a natural topology with respect to which it is homeomorphic to Euclidean space of dimension $6(\text{genus } M) - 6$.

In these lectures we shall outline a proof that $\mathcal{T}(M)$ as the natural structure of a smooth C^∞ finite-dimensional manifold of dimension $6(\text{genus } M) - 6$ and that this manifold is diffeomorphic to Euclidean space $\mathbb{R}^{6(\text{genus } M) - 6}$.

An account with complete details can be found in the forthcoming book, "A Riemannian Approach to Teichmüller Theory" by the author and A.E. Fischer [9].

§ 2 The space $\mathcal{A}$ of Almost Complex structures

The space of all complex structures $\mathcal{C}$ is a bit difficult to "get one's hands on". In the next sections we shall successively pass to a series of equivalent models for $\mathcal{C}$, the final model helping us to understand the quotient space $\mathcal{T}(M)$.

A C^∞ (1,1) tensor H is a mapping such that for each $x \in M$, $H_x : T_xM \longrightarrow T_xM$ is a linear mapping from the tangent space T_xM to itself and such that $x \longrightarrow H_x$ is C^∞ smooth. The space of all C^∞ (1,1) tensors forms an infinite dimensional Fréchet space. The space of almost complex structures $\mathcal{A}$ is the subset of all C^∞ (1,1) tensors J, such that for each $x \in M$, $J_x^2 = -\mathrm{id}_x$, $\mathrm{id}_x : T_xM \longrightarrow T_xM$ the identity map and secondly for each tangent vector $X_x \in T_xM$, X_x, J_xX_x is an oriented basis for T_xM.

One can naturally ask what sort of structure A has. Is it a manifold? A variety? The answer to this question is provided by the following theorem.

Theorem 1. A is a C^∞ smooth Fréchet manifold. The tangent space to A at a $J_0 \in A$ is the subspace of all C^∞ (1,1) tensors H satisfying the algebraic relation $HJ_0 = -J_0H$.

Proof. Let $C^\infty(T_1^1(M))$ denote the space of $C^\infty(1,1)$ tensors. Define a mapping $\Phi : C^\infty(T_1^1(M)) \longrightarrow C^\infty(T_1^1(M))$ by

$$\Phi(H) = H^2$$

Then $A = \Phi^{-1}(-\mathrm{id})$, where $-\mathrm{id}_x : T_xM \longrightarrow T_xM$ is the identity mapping. If (i) A were modeled on a Banach space (C^r or Sobolev space of tensors H^s) where the inverse function theorem applied and (ii) if $-\mathrm{id}$ were a regular value for Φ then by the implicit function theorem we would be able to conclude that A was a manifold (in fact a submanifold of $C^\infty(T_1^1(M))$ and moreover that the tangent space $T_{J_0}A$ is given by the kernel of the differential of Φ, namely

$$D\Phi_J(H) = HJ + JH$$

or

$$(2.1) \qquad T_{J_0}A = \{H \mid HJ = -JH\}.$$

Although neither (i) nor (ii) is valid the tangent space to A at J_0 is given by 2.1. In fact we can write down explicit global coordinates for A discovered by Uwe Abresch and Arthur Fischer. For fixed J_0, define

$$\varphi : T_{J_0} A \longrightarrow A$$

by

(2.2) $$\varphi(H) = (I + H) J_0 (I + H)^{-1} .$$

Formula (2.2) makes it clear that φ maps $T_{J_0} A$ to A. Since J_0 anticommutes with H we can rewrite φ symbolically as

$$\varphi(H) = J_0 \cdot \frac{I-H}{I+H}$$

which shows that φ is a linear fractional transformation. What is more surprising is that φ has an "algebraic inverse". We leave it to the reader to verify that if $\psi = \varphi^{-1}$,

$$\psi(J) = (J - J_0)(J + J_0)^{-1}$$

$$= \frac{J-J_0}{J+J_0}$$

In two dimensions it follows that $J^2 = -\mathrm{id}$, $J_0^2 = -\mathrm{id}$ implies that $J + J_0$ fails to be invertible only when $J = -J_0$. However $J = -J_0$ (because of the orientation condition) is <u>not</u> in A. From this we can conclude that φ is a <u>global</u> chart for A which concludes theorem 1.

Before going further in developing Teichmüller's theory we wish to make one further remark about the manifold A. One can ask if this manifold of almost complex structures is, in fact, a complex manifold.

We can begin by asking the simpler question of whether A is an almost complex manifold? The answer is yes, in a very natural sense.

Define, for each $J \in A$, a linear map $\Xi_J : T_J A \longrightarrow T_J A$ by

(2.3) $$\Xi_J(H) = JH .$$

It follows immediately that

$$\Xi_J^2 = -I ,$$

$-I_J : T_J A \longrightarrow T_J A$ the identity mapping. Thus A is an <u>almost complex manifold</u> with <u>almost complex structure</u> Ξ. A will be a <u>complex manifold</u> if there is a collection of coordinate charts (φ_i, U_i), U_i open in $T_{J_0} A$ for some fixed J_0 such that $\varphi_i(U_i)$ cover A and $\varphi_i^*(\Xi) = \Xi_{J_0}$ the fixed almost complex structure on $T_{J_0} A$. The surprising observation of Abresch-Fischer is that our global coordinate chart φ given in (2.2) satisfies

$$\varphi^*(\Xi) = \Xi_{J_0}$$

and thus A is a complex manifold.

We now return to the relation between A and the study of Teichmüller's moduli space.

Let $c \in C$ be a complex structure. Then to each such c we can associate an almost complex structure $J \in A$ as follows. Let $x_0 \in M$ and $\psi : U \to \mathbb{R}^2$ $U \subset M$ an open neighborhood of x_0 in M, ψ a complex coordinate chart for M. Now $\mathbb{R}^2$ has a natural almost complex structure $\hat{J}$ given by the matrix $\begin{pmatrix} 0 & -1 \\ 1 & 0 \end{pmatrix}$. For $x \in U$ define $J_x : T_x M \to T_x M$ by

$$J_x = d\varphi_x^{-1}\hat{J}d\varphi_x .$$

On the surface it would appear that J_x depends on the choice of coordinate mapping φ. However, we leave it as an easy exercise for the reader to verify that if ψ is any other complex coordinate mapping for a neighborhood of x_0,

$$J_x = d\psi_x^{-1}\hat{J}d\psi_x .$$

Thus $\{J_x\}_{x\in M}$ is a globally defined almost complex structure on M. This defines a mapping

(2.4) $\quad \mathcal{C} \longrightarrow \mathcal{A} , \quad c \longrightarrow J$

Our next observation is that the diffeomorphism group $\mathcal{D}$ acts on $\mathcal{A}$ in a natural way, namely, if $f \in \mathcal{D}$

$$(f,J) \longrightarrow f^*J = df_x^{-1} J_{f(x)} df_x .$$

Moreover $c \longrightarrow J$ has the property that it is $\mathcal{D}$ equivariant, namely if $c \longrightarrow J$, $f^*c \longrightarrow f^*J$.

Later we shall see that this map has a natural inverse and is therefore bijective. Thus in order to study $\mathcal{T}(M) = \mathcal{C}/\mathcal{D}_0$ it suffices to study $\mathcal{A}/\mathcal{D}_0$.

Finally we remark that the almost complex structure Ξ is $\mathcal{D}$-invariant and leave this verification for the reader. This last observation is important in realizing the quotient space, Teichmüller's space $\mathcal{T}(M) \approx \mathcal{A}/\mathcal{D}_0$ as a "complex manifold".

§ 3 The space of Riemannian Metrics $\mathcal{M}$

Let S_2 be the space of all $(0,2)$ tensors on M. Thus $h \in S_2$ iff for each $x \in M$, $h(x) : T_xM \times T_xM \longrightarrow R$ is a symmetric bilinear form. Let $\mathcal{M} = \{g \in S_2 \mid g(x) > 0$ for each $x \in M\}$. Thus of $v \in T_xM$, $g(x)(v,v) > 0$ if $v \neq 0$. Clearly $\mathcal{M} \subset S_2$ is open and thus a submanifold. The diffeomorphism group $\mathcal{D}$ also acts on $\mathcal{M}$, namely $(f,g) \longrightarrow f^*g$ where

$$(f^*g)(x)(u,v) = g(f(x)(df_xu, df_xv)$$

$u,v \in T_xM$.

There is a very natural mapping from $\mathcal{M}$ to $\mathcal{A}$ described as follows. For $g \in \mathcal{M}$, $v \in T_xM$, define $J_x : T_xM \to T_xM$ by J_xv is the rotation of v by 90° in the "counter-clockwise direction". Given $g \in \mathcal{M}$ and the orientation on M this is well defined. This gives us a map $\Psi : \mathcal{M} \longrightarrow \mathcal{A}$.

The problem with this definition is that it is implicit as opposed to explicit. As analysts we would like to be able to differentiate mappings, even in the infinite-dimensional context Fortunately we can describe the mapping Ψ explicitly, namely

$$(3.1) \qquad \Psi(g) = -g^{-1}\mu_g$$

where $\mu_g(x) : T_xM \times T_xM \longrightarrow R$ is the unique antisymmetric two form, the volume element associated to g and the given orientation of M. More precisely, by $\Psi(g) = J$ we mean

$$(3.1)' \qquad -g(x)(J_xu,v) = \mu_g(x)(u,v)$$

for all $u, v, \in T_xM$.

It is not to hard to check that $\Psi : \mathcal{M} \longrightarrow \mathcal{A}$ and is $\mathcal{D}$-equivariant, namely

$$\Psi(f^*g) = f^*\Psi(g) \ .$$

However Ψ is <u>not</u> a bijection and this is not difficult to see. Let $p : M \longrightarrow R^+$ be a strictly positive C^∞ function. Then $p \cdot g \in \mathcal{M}$ if $g \in \mathcal{M}$ $\{(p \cdot g)(x)(u,v) = p(x)g(x)(u,v)\}$. Since $\mu_{p \cdot g} = p \cdot \mu_g$ one sees immediately from (3.1)' that

$$\Psi(pg) = \Psi(g)$$

Let $\mathcal{P}$ be the space of all C^∞ positive functions. One can check that $\mathcal{P}$ acts freely and properly on $\mathcal{M}$, the action given by the map

$$(p,g) \longmapsto p \cdot g \ .$$

If we were in the C^∞ or Sobolev H^s class of metrics and functions it would follow immediately that the quotient space, say $\mathcal{M}^s/\mathcal{P}^s$ is a manifold. However in the C^∞ class of metrics $\mathcal{M}$ and C^∞ positive functions $\mathcal{P}$ this result is still true, namely that $\mathcal{M}/\mathcal{P}$ is a C^∞ Fréchet manifold. Before we state this as part of the main theorem of this section let us see what the tangent space of this manifold must be. First, if $h \in S_2$ and $g \in \mathcal{M}$ we can define the trace of h with respect to g, by

$$\text{(3.2)} \qquad \operatorname{trace}_g h = \operatorname{tr}_g h = \operatorname{trace} H$$

where H is the $(1,1)$ tensor associated to h via the metric g via the relation

$$(3.3) \qquad g(x)(H_x u,v) = h(x)(u,v)$$

$x \longmapsto (tr_g h)_x$ is a C^∞ function on M. Now let $h \in S_2$. Then

$$(3.4) \qquad h = (h - \tfrac{1}{2}(tr_g h) \cdot g) + (\tfrac{1}{2} tr_g h) \cdot g$$

$$= h^T + \lambda \cdot g$$

where $tr_g h^T = 0$.

Thus every $(0,2)$ tensor h can be decomposed as a trace free tensor plus a multiple of the metric g.

Consider now the map

$$p \longmapsto p \cdot g$$

for fixed g. The image of this mapping is the orbit of the group P through the point g. The tangent space of this orbit is clearly the set of $\lambda \cdot g$ where $\lambda : M \longrightarrow R$ is a C^∞ function. From this observation and the previous remarks we can state the following

<u>Theorem 2.</u> The quotient space M/P is a C^∞ Fréchet manifold whose tangent space at an equivalence class $[g] \in M/P$ can be identified with those $h \in S_2$ such that $tr_g h = 0$.

Moreover we have

<u>Theorem 3.</u> The map $\Psi : M \longrightarrow A$ induces a C^∞ D-equivariant diffeomorphism $\Psi : M/P \longrightarrow A$. The conclusion of this section

is that:

In order to understand the structure of the space $\mathcal{T}(M) \approx \mathcal{A}/\mathcal{D}_0$ it suffices to understand the structure of the double quotient space $\mathcal{T}(M) \approx \mathcal{M}/\mathcal{P}/\mathcal{D}_0$.

§ 4 The Manifold $\mathcal{M}_{-1}$ of Metrics of Negative Constant Curvature Minus One

In this section and later we build upon a classical and important result of Poincaré:

Theorem 4. Let M be an oriented surface of (genus M) > 1 . Then given any metric $g \in \mathcal{M}$ there exists a unique $\lambda : M \longrightarrow R$ such that the Gauss or scalar curvature of λg , $R(\lambda g)$ is the constant function -1 .

Although Poincaré probably proved this theorem using the uniformization theorem for Riemann surfaces, today it is possible to give a straightforward proof using the calculus of variations. If g is given, we write $\lambda = e^u$. The condition that $R(e^u g) = -1$ reduces to a quasi-linear second order p.d.e. whose highest order term is just the Laplacian of u . This p.d.e. is the Euler-Lagrange equation of a simple-variational problem which can be shown to have a regular minimum. A maximum principle argument ensures that the solution to the p.d.e. is unique. For details see [10].

The Gauss curvature of a metric g is a function on M . Thus the Gauss curvature itself can be thought of as a smooth function $R : \mathcal{M} \longrightarrow \mathcal{F}$, where $\mathcal{F}$ is the space of C^∞ functions on M . If we were working in the Sobolev class of H^s metrics $\mathcal{M}^s$ and H^s functions $\mathcal{F}^s$, R would then be a C^∞ map from $\mathcal{M}^s$ to $\mathcal{F}^{s-2}$.

This important result of Poincaré suggests that $\mathcal{M}_{-1}$ is bijectively equivalent to $\mathcal{M}_{-1}$. We begin by asking whether $\mathcal{M}_{-1}$ is a manifold. This is answered by

Theorem 5. The space $\mathcal{M}_{-1}$ of metrics of constant negative Gauss curvature is a smooth C^∞ Fréchet submanifold of $\mathcal{M}$.

Proof. We sketch a proof in the Sobolev class $\mathcal{M}^s$ of metrics. This result follows immediately from the implicit function theorem. The result in the C^∞ case involves slightly more work. That $\mathcal{M}^s_{-1}$ is a manifold would follow if we could show that -1 is a regular value for the map

$$R : \mathcal{M}^s \longrightarrow \mathcal{F}^{s-2} .$$

Now the derivative formula for

$$DR(g) : S^s_2 \longrightarrow \mathcal{F}^{s-2}$$

$S^s_2 \approx T_g\mathcal{M}^s$ the H^s smooth $(0,2)$ tensors on M, is well known. In the case $R(g) \equiv -1$ it is given by the formula

$$DR(g)h = -\Delta(\mathrm{tr}_g h) + \delta_g\delta_g h + \frac{1}{2}(\mathrm{tr}_g h) \tag{4.1}$$

where Δ is the Laplace-Betrami operator on functions, $\mathrm{tr}_g h$ is the trace of h with respect to g, and $\delta_g\delta_g h$ is the function on M defining the double covariant divergence of h with respect to g [see Eisenhart [4]].

We may conclude that $\mathcal{M}^s_{-1}$ is a submanifold of $\mathcal{M}^s$ with tangent space Kernel $DR(g)$ if we can show that the operator $h \longmapsto DR(g)h$ is onto $\mathcal{F}^{s-2}$. The simplest way to see this surjectivity is not by considering __all__ variations h but by considering the subspace of those h of the form $\rho \cdot g$ where $\rho \in F^s$. Then $tr_g(\rho g) = 2\rho$ and $\delta_g \delta_g(\rho g) = \Delta\rho$, again Δ the Laplace-Beltrami. Thus if $h = \rho g$,

(4.2) $$DR(g)h = -2\Delta\rho + \Delta\rho + \rho$$

$$= -\Delta\rho + \rho \ .$$

Now it is completely standard in elliptic theory that $\rho \longmapsto -\Delta\rho + \rho$ as a map from $F^s \longrightarrow F^{s-2}$ is surjective. This concludes the proof of theorem 5. But we get more:

__Theorem 6.__ The tangent space to the submanifold $M_{-1} \subset S_2$ consists of those (0,2) tensors h such that

(4.3) $$-\Delta(tr_g h) + \delta_g \delta_g h + \frac{1}{2}(tr_g h) = 0 \ .$$

Consider now the natural quotient map $\pi : M \longrightarrow M/P$. Let π_{-1} be the restriction of π to the submanifold M_{-1}. We can now prove

__Theorem 7.__ The map $\pi_{-1} : M_{-1} \longrightarrow M/P$ is a D-equivariant diffeomorphism of Fréchet manifolds.

__Proof.__ First we should remark that D acts on M_{-1}; i.e. the D action on M preserves M_{-1}. This follows from the relation that

(4.4) $$R(f^*g) = f^*R(g) = R(g) \circ f \ .$$

Thus if $R(g) \equiv -1$, then $R(f^*g) \equiv -1$. That π_{-1} is $\mathcal{D}$-equivariant follows from the fact that $\pi : \mathcal{M} \longrightarrow \mathcal{M}/\mathcal{P}$ is naturally $\mathcal{D}$-equivariant.

From Poincaré's theorem 4 it follows that π_{-1} is a bijection. We must show that π_{-1} is a diffeomorphism, or that $D\pi_{-1} : T_g\mathcal{M}_{-1} \longrightarrow T_{[g]}\mathcal{M}/\mathcal{P}$ is an isomorphism of Fréchet spaces. From theorem 2 we know that $T_{[g]}\mathcal{M}/\mathcal{P}$ can be identified with those elements of S_2 which are trace-free with respect to g . Moreover $D\pi_g(h) = h^T = \{h-(\frac{1}{2}tr_g h)g\}$, h^T the trace-free part of h . From this it immediately follows that

$$(4.5) \qquad D\pi_{-1}(h) = h^T .$$

Now suppose $D\pi_{-1}(h) = 0$. This means that $h = \rho g,\ \rho \in \mathcal{F}$. But $DR(g)h = 0$ which implies as an 4.2 that

$$-\Delta\rho + \rho = 0$$

or that $\rho = 0$. Thus $D\pi_{-1}$ is injective. Now let $v \in \mathcal{F}$. To show $D\pi_{-1}$ is surjective we must produce an $h \in T_g\mathcal{M}_{-1}$ such that $D\pi_{-1}(h) = v$. Let $h = \rho g$. Then it suffices to find a ρ such that

$$-\Delta\rho + \rho = v .$$

But $\rho \longrightarrow -\Delta\rho + \rho$ is surjective which completes the proof of theorem 7.

The conclusion of this section is that in order to understand the structures of $\mathcal{M}/\mathcal{P}/\mathcal{D}_0 \cong \mathcal{A}/\mathcal{D}_0 \cong \mathcal{C}/\mathcal{D}_0$ it suffices to understand the structure of $\mathcal{M}_{-1}/\mathcal{D}_0$. Our intuition is now very clear. We have an infinite-dimensional Lie group $\mathcal{D}_0$ acting on a smooth C^∞ manifold M_{-1} and we wish to understand the quotient space $M_{-1}/\mathcal{D}_0$. Our first objective is to identify the tangent space to the quotient. This we do in section 6.

§ 5 Conformal Coordinates and the D-equivariant Equivalence Between M_{-1} and C, and C and A

Let $x \in M$ and $g \in M$ be arbitrary. Then it is a standard fact that we can find a diffeomorphism $\psi : V \longrightarrow U$ where V is a neighborhood of 0 in $\mathbb{R}^2$ and U is a neighborhood of x in M such that the pull-back of the metric g

$$(5.1) \qquad (\psi^* g)_{ij} = \lambda \delta_{ij}$$

where δ_{ij} is Kronecker's delta; i.e. the pull-back of the metric g to a metric on $V \subset \mathbb{R}^2$ is given by a positive function times the standard Euclidean metric. The set of all $\{\psi^{-1}, U\}$ naturally forms a complex structure for M .

What this means is the following: If $g \in M_{-1}$ a conformal coordinate system yields a complex structure $c \in C$.

One can readily see that this is a $\mathcal{D}$-equivariant mapping which is the inverse of the composition of the maps

$$c \longrightarrow \tau \longrightarrow g \in M_{-1}$$

developed in the last sections. Thus the existence of conformal coordinate systems provide an inverse for the map $c \longrightarrow \tau$, $C \longrightarrow A$ introduced in § 2.

§ 6 The L_2-Decomposition of the Tangent Space to $\mathcal{M}_{-1}$

Let $g \in \mathcal{M}_{-1}$ be fixed and consider the orbit of the diffeomorphism group $\mathcal{D}$; namely the manifold

$$\mathcal{O}_{\mathcal{D}}(g) = \{f^{\alpha}g / f \in D\} .$$

What is the tangent space to this orbit? Let f_t be a smooth 1-parameter group of diffeomorphisms with $-\varepsilon < t < \varepsilon$, $\varepsilon > 0$, $f_0(x) = x$, $\left.\frac{df}{dt}\right|_{t=0} = X$, a vector field on M . A tangent vector $\beta \in T_g\mathcal{O}_{\mathcal{D}}(g)$ will be given by

$$\text{(6.1)} \qquad \frac{d}{dt}\left\{f_t^* g\right\}_{t=0} = \beta .$$

But (6.1) is a well-known object in differential geometry, the Lie derivative of g with respect to the vector field X . In a local coordinate system where g is given by g_{ij}, we can write this Lie derivative $L_X g$ as

$$\text{(6.2)} \qquad (L_X g)_{ij} = X^k \frac{\partial g_{ij}}{\partial x^k} + g_{kj}\frac{\partial x^k}{\partial x^i} + g_{ik}\frac{\partial x^k}{\partial x^j} .$$

Summarizing we have:

Theorem 8. The tangent space to the orbit $\mathcal{O}_{\mathcal{D}}(g)$ is the set of all Lie derivatives $L_X g$, X a vector field on M .

If we are to have a natural candidate to the quotient space $\mathcal{M}_{-1}/\mathcal{D}_0$ we should produce a natural complement, say an orthogonal complement to the tangent space $T_g\mathcal{O}_{\mathcal{D}}(g)$. But "orthogonal complement" necessitates the existence of a metric. Therefore we ask:

Is there a "natural" metric on the space of all metrics $\mathcal{M}$, where by natural we mean that the diffeomorphism group should act as a group of isometries? The positive answer was well known to the physicists and general relativists for a long time, namely the so called L_2-metric on the space of metrics which is defined as follows. Let $h,k \in T_g\mathcal{M}_{-1}$ and let H,K be the (1,1) tensors $\{g(x)(H_x u,v) = h(x)(u,v)\}$ corresponding to h,k via the metric g. Define

(6.3) $$\langle\langle h,k\rangle\rangle_g = \int_M \operatorname{trace}(HK)\,d\mu_g \, .$$

Here $x \longmapsto \operatorname{trace}(H_x K_x)$ is a C^∞ function on M integrated with respect to the volume element μ_g. Clearly

$$\langle\langle\, ,\rangle\rangle : T_g\mathcal{M}_{-1} \times T_g\mathcal{M}_{-1} \longrightarrow R \, .$$

<u>Theorem 9.</u> $\mathcal{D}$ acts as a group of isometries for $\langle\langle\, ,\rangle\rangle$.

<u>Proof.</u> For $f \in \mathcal{D}$ consider the map

(6.4) $$\theta_f(g) = f^*g \, .$$

Since $g \longmapsto f^*g$ is " linear" in g it follows that the derivative

(6.5) $$D\theta_f(g) : S_2 \longrightarrow S_2$$
$$D\theta_f(g)(h) = f^*h \, .$$

θ_f is an isometry if

$$(6.6) \qquad \langle\langle D\theta_f(g)(h), D\theta_f(g)k\rangle\rangle_{\theta_f(g)} = \langle\langle h,k\rangle\rangle_g \ .$$

But the left hand side of (6.6) is equal to

$$\int_M \operatorname{trace}(f^* H f^* K)_x \, d\mu_{f^* g}$$

$$= \int_M \operatorname{trace}(HK)_{f(x)} \, d\mu_{f^{\alpha} g}$$

which by the ordinary change of variables formula is equal to

$$\int_M \operatorname{trace}(HK)_x \, d\mu_g$$

which proves theorem 9.

The next question is whether we can identify the orthogonal complement of those $L_X g \in T_g M_{-1}$ with respect to this metric. The first step in this direction is given by the L_2-decomposition of symmetric tensors given in a famous splitting theorem whose first complete proof was given by Berger and Ebin.

Theorem 10 (Berger-Ebin). Let $h \in S_2$ and $g \in M$. Then h can be uniquely decomposed as

$$(6.7) \qquad h = h^{o} + L_X g$$

where h^{o} is a divergence free $(0,2)$ tensor; i.e.

$$\delta_g h^{o} = 0$$

where in a local coordinate system

$$(6.8) \qquad (\delta_g h)_i = \frac{1}{\sqrt{g}} \frac{\partial}{\partial x^j} (h_i^j \sqrt{g}) - \frac{1}{2} g^{kl} h_l^i \frac{\partial g_{jk}}{\partial x^i}$$

where $\sqrt{g} = \det(g_{ij})$ the determinant of the g_{ij}, g^{kl} is the matrix of the inverse of $\{g_{ij}\}$, and finally $h^j_i = g^{jk}h_{ki}$; and in local coordinates h is given by $\{h_{ki}\}$. Moreover the decomposition (6.7) is L_2-orthogonal.

We are now in a position to determine the orthogonal complement in $T_g M_{-1}$ of the space of Lie derivatives $L_X g$, the tangent space to the orbit of $\mathcal{D}$. Let $g \in M_{-1}$ and suppose that

$$h \in T_g M_{-1} \{-\Delta(\mathrm{tr}_g h + \delta_g \delta_g h + \tfrac{1}{2}(\mathrm{tr}_g h) = 0\} .$$

Then by Berger-Ebin,

$$h = h^{\circ} + L_X g .$$

But, since $L_X g \in T_g M_{-1}$ it follows that

$$0 = -\Delta(\mathrm{tr}_g(L_X g)) + \delta_g \delta_g (L_X g) + \tfrac{1}{2}(\mathrm{tr}_g L_X g) .$$

Thus if $h \in T_g M_{-1}$, $h = h^{\circ} + L_X g$ then necessarily

$$-\Delta(\mathrm{tr}_g h^{\circ}) + \delta_g \delta_g h^{\circ} + \frac{1}{2}\,\mathrm{tr}_g h^{\circ} = 0 .$$

Since $\delta_g h^{\circ} = 0$, $\delta_g \delta_g h^{\circ} = 0$, whence we see that

$$-\Delta(\mathrm{tr}_g h^{\circ}) + \frac{1}{2}\,\mathrm{tr}_g h^{\circ} = 0$$

if $\rho = \mathrm{tr}_g h^{\circ}$, then

$$-\Delta\rho + \rho = 0$$

which immediately implies (just integrate by parts) that $\rho = 0$.

Our conclusion is the following: Let $h \in T_g M_{-1}$. Then

$$h = h^o + L_X g$$

where h^o is both divergence-free and trace-free. For clarity it would be better to denote such a trace-free, divergence-free two-tensor by h^{TT} and the set of all such trace-free and divergence-free two-tensors with respect to g by $S_2{}^{TT}(g)$. Paraphrasing our result we see that

Theorem 11. Any $h \in T_g M_{-1}$ can be uniquely decomposed as

$$h = h^{TT} + L_X g$$

where $h^{TT} \in S_2{}^{TT}(g)$, and this decomposition is L_2-orthogonal.

We should remark that in general, although $L_X g$ is unique, X need not be uniquely determined. It will be determined only up to a Killing field Y, $L_Y g = 0$. If $g \in M_{-1}$ there are no non-zero Killing fields ($L_Y g = 0$ implies $Y = 0$).

Back to Teichmüller space! If $M_{-1}/\mathcal{D}_0$ is a manifold then the dimension of the space of $S_2^{TT}(g)$ must be finite and the same dimension for all $g \in M_{-1}$. How can we compute this dimension? Let us see what it means for h^{TT} to belong to $S_2{}^{TT}(g)$ in a local conformal coordinate system for M (recall that such a conformal coordinate system is also a complex coordinate system). In such a system we can write h^{TT} in classical notation

$$h^{TT} = u\, dx^2 - u\, dy^2 - 2\, v dx dy$$

u,v local functions on the coordinate neighborhood.

We have already used the fact that h^{TT} is trace-free by writing the coefficient of dy^2 as the negative of that of dx^2. What does it mean for h^{TT} to be divergence-free? Using formula 6.8 the reader may verify that $\delta_g h^{TT} = 0$ reduces to the system of equations

$$\frac{\partial u}{\partial x} = \frac{\partial v}{\partial y}$$

$$\frac{\partial u}{\partial y} = -\frac{\partial v}{\partial x},$$

that is, the Cauchy-Riemann equations! Thus, formally

$$h^{TT} = \mathrm{Re}\{(u+iv)(dx+idy)^2\} \tag{6.9}$$

$$= \mathrm{Re}\{\xi(z)\,dz^2\}$$

where $\xi(z)$ is a local holomorphic function.

Thus h^{TT} is the real part of a complex valued symmetric tensor which in conformal coordinates is given by a holomorphic function $\xi(z)$. This is the definition of a holomorphic quadratic differential on M with respect to the complex structures $c(g)$ arising from the metric g.

We are now back to Teichmüller's original observation that as a consequence of the theorem of Riemann-Roch, one can compute the dimension of the space of holomorphic quadratic differentials to be $6(\text{genus } M)-6$, exactly the number that Riemann had conjectured was the dimension of $R(M)$, the Riemann space of moduli {cf. § 2}. As a corollary we obtain

Corollary 12. For any $g \in M_{-1}$

$$\dim S_2^{TT}(g) = 6(\text{genus } M) - 6 .$$

Paraphrasing theorem 11 we have

Theorem 11'. Any $h \in T_g M_{-1}$ can be uniquely decomposed as

$$h = \mathrm{Re}(\xi(z)dz^2) + L_X g$$

where $\xi(z)dz^2$ is a holomorphic quadratic differential.

Thus if Teichmüller space is a manifold it must be of dimension $6(\text{genus } M) - 6$! In the next section we construct a coordinate chart for Teichmüller space.

§ 7 Teichmüller's Moduli Space is a Smooth C^∞ Manifold of Dimension 6(genus M) - 6

Given $g \in M_{-1}$ we would like to construct a submanifold $S_g \subset M_{-1}$ through g transverse to the orbit $O_{\mathcal{D}}(g)$ of $\mathcal{D}$ through g.

Let $h^{TT} \in S_2^{TT}(g)$ be "small". Then $g + h^{TT} \in M$. By Poincaré's theorem 4 we can find a unique function $\lambda(h^{TT}) : M \longrightarrow R^+$ such that the Gauss curvature $R(\lambda(h^{TT})\{g + h^{TT}\}) \equiv -1$. Moreover $h^{TT} \longmapsto \lambda(h^{TT})$ is smooth and $\lambda(0) = 1$, since $R(g) \equiv -1$. Consider the map

$$h^{TT} \overset{\Theta}{\longmapsto} \lambda(h^{TT})\{g + h^{TT}\} .$$

One sees that $D\Theta(0)h^{TT} = h^{TT}$. By the standard finite-dimensional implicit function theorem this implies that the image under Θ of a neighborhood of 0 in $S_2^{TT}(g)$ is a submanifold S_g of M_{-1} of dimension $6(\text{genus } M) - 6$. This "local slice" S_g is our candidate for a coordinate chart for $T(M) = M_{-1}/D_0$ (see figure below)

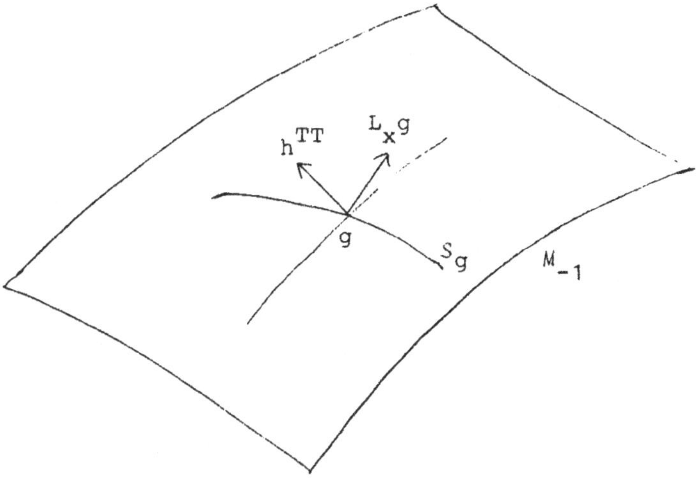

Up till now we have not distinguished between the action of D_0 and D . Through each point $\hat{g} \in S_g$ there is an orbit of D . It is possible that no matter how small S_g is chosen different points of S_g may correspond to the same orbits of D . This is exactly what happens. This is the reason why the Riemann space of moduli is not a manifold and Teichmüller space is, for if S_g is small enough each point $\hat{g} \in S_g$ corresponds to a unique orbit of D_0 , and thus locally, points in the quotient space may be uniquely identified with points in the slice S_g . This gives the C^∞ manifold structure to $T(M)$. The fact that each point $\hat{g} \in L_g$ corresponds to a unique orbit of D_0 follows from two theorems. The first, due to Ebin-Palais says that the action of D on M is proper. We state this as

Theorem 12 (Ebin-Palais). $\mathcal{D}$ acts properly on M, i.e. if $f_n \in \mathcal{D}$ and $g_n \in M$ are sequences such that $f_n^* g_n \longrightarrow g_1$ and $g_n \longrightarrow g_0$ then there is a subsequence of f_n which converges $f_n \longrightarrow f$, to some $f \in D$. For a proof see [2]

Theorem 13. $\mathcal{D}_0$ acts freely on M_{-1}.

Remark. This theorem is definitely not true if $\mathcal{D}_0$ is replaced by $\mathcal{D}$.

Proof. Suppose $f \in \mathcal{D}_0$ and $f^*g = g$ i.e. f is an isometry for g. If $c(g)$ is the unique complex structure associated to g it follows that $f^*c(g) = c(g)$. This says that $f : (M,c(g)) \longrightarrow (M,c(g))$ is a holomorphic self-map.

Let us suppose that f is not the identity. Then since f is holomorphic it follows that the fixed point set of f is isolated and therefore finite in number. Since f is holomorphic the Lefschetz fixed point index $\Lambda(f) > 0$. Since f is homotopic to the identity

$$\Lambda(f) = 2(1\text{-genus } M) < 0$$

the Euler characteristic of M. This is a contradiction. Therefore f must be the identity, which finishes theorem 13.

Theorem 14. If S_g is small enough each point $\hat{g} \in S_g$ corresponds to a unique orbit of $\mathcal{D}_0$.

Proof. Suppose not. Then there exists a sequence $f_n \in \mathcal{D}_0$ and $g_n \in S_g$ with $g_n \longrightarrow g$ and $f_n^* g_n \longrightarrow g$ with all f_n outside a fixed neighborhood of the identity map of M. By the Ebin-Palais lemma we may conclude that there exists a subsequence of the f_n, say f_n such that $f_n \longrightarrow f$. Then necessarily $f^*g = g$, $f \neq id$ and f homotopic to the identity contradicting theorem 13.

Thus a neighborhood of $[g]$ in $M_{-1}/\mathcal{D}_o$ can be identified with S_g which provides a coordinate chart for the quotient $\mathcal{T}(M) = M_{-1}/\mathcal{D}_o$. It is a standard exercise that the transition maps for these coordinate mappongs are C^∞ smooth. Thus we may conclude this section with

Theorem 15. Teichmüller's space $\mathcal{T}(M) \cong M_{-1}/\mathcal{D}_o \cong M/P/\mathcal{D}_o \cong \cong A/D_o \cong C/\mathcal{D}_o$ has the natural structure of a smooth C^∞ manifold of dimension $6(\text{genus } M) - 6$.

In the next section we shall see why Teichmüller's space must be diffeomorphic to Euclidean space.

§ 8 Teichmüller's Space is Diffeomorphic to $\mathbb{R}^{6(\text{genus } M) - 6}$

Our main objective in this section is to indicate that Dirichlet's energy on Teichmüller space is a proper smooth C^∞ function with one non-degenerate minimum point.

Let $g_o \in M_{-1}$ be a fixed metric and let $g \in M_{-1}$ be arbitrary. Furthermore for simplicity of exposition we assume (without loss of generality) that (M,g_o) is isometrically embedded in some Euclidean space of dimension K.

Let $S : (M,g) \longrightarrow (M,g_o) \subset \mathbb{R}^K$ be a smooth map. Then $S = (S^1,\dots,S^K)$. Define Dirichlet's energy of S by

$$(8.1) \qquad E(g,S) = \frac{1}{2} \sum_{j=1}^{K} \int_M g(x)(\nabla_g S^j, \nabla_g S^i)\, d\mu_g$$

$\nabla_g S^i$ the gradent of the jth component function S^j with respect to the metric g .

We now have a basic result towards which many people have contributed including Eells, Sampson, Schoen, Yau and Jost.

Theorem 16. For each fixed $g \in \mathcal{M}_{-1}$ there exists a unique critical point (a minimum point) $S(g) : M \longrightarrow M$ of E which is homotopic to the identity. Moreover $S(g) \in \mathcal{D}_0$ and depends smoothly on g . $S(g)$ is a harmonic map from (M,g) to (M,g_0) .

Dirichlet's energy has an invariance property under the action of D , namely

$$(8.2) \qquad E(f^*g, f^*S) = E(g,S)$$

where $f^*(S) = S \circ f$. Thus E as a function of two variables is D-invariant. Define the function $\tilde{E} : \mathcal{M}_{-1} \longrightarrow R$ by

$$(8.3) \qquad \tilde{E}(g) = E(g,S(g)) \ .$$

Theorem 17. The function $\tilde{E}$ is $\mathcal{D}_0$ invariant (but not $\mathcal{D}$ invariant!).

Proof. $\tilde{E}(f^*g) = E(f^*g, S(f^*g))$.

Now $S(f^*g)$ is the unique harmonic map from (M,f^*g) to (M,g_0) homotopic to the identity. But $f^*S(g) = S(g)\circ f$ is also a harmonic map from (M,f^*g) to (M,g_0) . If $f \in \mathcal{D}_0$, $S(g)\circ f \in \mathcal{D}_0$ and therefore by uniqueness

$$S(f^*g) = f^*S(g) \ .$$

By (8.2) we may conclude that

$$\widetilde{E}(f^*g) = \widetilde{E}(g),\ f \in \mathcal{D}_0$$

concluding theorem 17.

Thus the smooth map $\widetilde{E}$ passes to a smooth map

$$\widetilde{E} : \mathcal{T}(M) \longrightarrow R .$$

Based on results of Schoen-Yau and Jost, the author observed that:

Theorem 18. $\widetilde{E} : \mathcal{T}(M) \longrightarrow R$ is a smooth proper map; i.e. the inverse image of a compact set is compact.

This result, the details of which we cannot go into in these two lectures is (in the opinion of this author) the analytical heart of Teichmüller theory.

To state our final result we would like to have a minor diversion and discuss a natural Riemannian metric on $\mathcal{T}(M)$, the Weil-Petersson metric.

Classical Teichmüller theory realizes the holomorphic quadratic differentials as the co-tangent space to Teichmüller's space. Around 1952, André Weil suggested to Lars Ahlfors a metric on Teichmüller space. Petersson, had earlier introduced a bilinear pairing on modular forms of weight k. Holomorphic quadratic differentials are a particular case of modular forms of weight 2. If ξ and η are holomorphic quadratic differentials on $(M,c(g))$, $g \in \mathcal{M}_{-1}$ and $c(g)$ the associated complex structure consider the expression

$$(\xi(z)dz^2)\cdot\overline{(\eta(z)dz^2)}$$

$$= \xi(z)\bar{\eta}(z)|dz|^4 .$$

In conformal coordinates the metric g can be written as $\lambda|dz|^2$. Thus the expression

$$\frac{\xi(z)\bar{\eta}(z)|dz|^4}{\lambda^2|dz|^4}$$

$$= \frac{\xi(z)\bar{\eta}(z)}{\lambda^2}$$

is invariant; i.e. independent of the choice of coordinate system and is thus a function on M. The metric on $T(M)$ suggested by Weil is the Petersson pairing, for $[g] \in T(M)$

$$<\xi,\eta>_{[g]} = \mathrm{Re}\int_M \frac{\xi(z)\overline{\eta(z)}}{\lambda^2}\, d\mu_g$$

$$= \mathrm{Re}\int_M \frac{\xi(z)\overline{\eta(z)}}{\lambda}\, |dz|^2 .$$

There is another way of obtaining a nice metric on $T(M) = M_{-1}/\mathcal{D}_0$. We have already seen that $\mathcal{D}$ acts as a group of isometries with respect to the L_2-metric $<< , >>$ on M_{-1}. Thus this metric passes to a metric $< , >$ on $M_{-1}/\mathcal{D}_0$ in the following way: Let $h,k \in T_{[g]}M_{-1}/\mathcal{D}_0$. Then there exists unique $\tilde{h},\tilde{k} \in T_gM_{-1}$, $\tilde{h},\tilde{k} \in S_2^{TT}(g)$ with $D\pi_{-1}(\tilde{h}) = h$, $D\pi_{-1}(\tilde{k}) = k$. Define

$$<h,k>_{[g]} = <<\tilde{h},\tilde{k}>>_g .$$

In fact that $\mathcal{D}$ acts as a group of isometries implies that $< , >$ is well defined on $T(M)$.

Theorem 19. Suppose $\tilde{h} = \mathrm{Re}\ \xi(z)dz^2$, $\tilde{k} = \mathrm{Re}\ \eta(z)dz^2$. Then

$$<\xi,\eta>_{[g]} = <h,k>_{[g]}$$

i.e. the Weil-Petersson metric is the quotient metric on $M_{-1}/\mathcal{D}_0$.

Let us return to Dirichlet's energy map $\tilde{E} : \mathcal{T}(M) \longrightarrow R$. A non-trivial calculation now yields

Theorem 20. $\tilde{E}$ has only one critical point, namely $[g_0]$, which is a minimum. The Hessian of $\tilde{E}$ at $[g_0]$ is given by the formula

$$D^2\tilde{E}[g_0](h,k) = <h,k>_{[g]}$$

i.e.the Weil-Petersson metric on $\mathcal{T}(M)$. Thus $[g_0]$ is a non-degenerate minimum point for $\tilde{E}$.

As an immediate consequence of theorems 18 and 20 we have

Theorem 21. Teichmüller's space is diffeomorphic to $\mathbb{R}^{6(\text{genus } M)-6}$.

Proof. Applying a simple case of the gradient deformation techniques of Morse theory we see that any n-manifold which admits a proper function which is bounded below and has only one non-degenerate minimum is diffeomorphic to Euclidean space. This ends the proof of Teichmüller's theorem.

Bonn, August 1987

References

[1] EARLE, C.J. and EELLS, J.: Deformations of Riemann Surfaces, Lecture Notes in Mathematics 103, Springer-Verlag (1969).

[2] EBIN, D.: The manifold of Riemannian metrics, Proc. Symp. Pure Math. AMS (15) 11-40 (1970).

[3] EELLS, J. and SAMPSON, J.H.: Harmonic mappings of Riemannian manifolds, Amer. J. Math. 86 (1964) 109-160.

[4] EISENHART, L.P.: Riemannian Geometry, Princeton Univ. Press (1966).

[5] FISCHER, A.E. and TROMBA, A.J.: On a purely Riemannian proof of the structure and dimension of the unramified moduli space of a compact Riemann surface. Math. Ann. 267 (1984) 311-345.

[6] FISCHER, A.E. and TROMBA, A.J.: On the Weil-Petersson metric on Teichmüller space, Trans. AMS 284 (1984), 319-335.

[7] FISCHER, A.E. and TROMBA, A.J.: Almost complex principle bundles and the complex structure on Teichmüller space, Crelle J. Band 252, pp. 151-160 (1984).

[8] FISCHER, A.E. and TROMBA, A.J.: A new proof that Teichmüller space is a cell, Trans. AMS (to appear).

[9] FISCHER, A.E. and TROMBA, A.J. "A Riemannian Approach to Teichmüller Theory" (Book) to appear.

[10] TOMI, F. and TROMBA, A.J.: Existence theorems for minimal surfaces of high genus in Euclidean space, Memoirs AMS (to appear).

[11] TROMBA, A.J.: On a natural algebraic affine connection on the space of almost complex structures and the curvature of Teichmüller space with respect to its Weil-Petersson metric, Manuscripta Math. vol. 56, Fas. 4, 475-497 (1986).

[12] TROMBA, A.J.: On an energy function for the Weil-Petersson metric; Manuscripta Math. (to appear).

[13] SHOEN, R. and YAU, S.T.: On univalent harmonic maps between surfaces, Inventiones mathematicae 44, 265-278 (1978).

Max-Planck-Institut für Mathematik
Bonn

and

Department of Mathematics
University of California
Santa Cruz

FONDAZIONE C.I.M.E.
CENTRO INTERNAZIONALE MATEMATICO ESTIVO
INTERNATIONAL MATHEMATICAL SUMMER CENTER

"Logic and Computer Science"

is the subject of the First 1988 C.I.M.E. Session.

The Session, sponsored by the Consiglio Nazionale delle Ricerche and the Ministero della Pubblica Istruzione, will take place under the scientific direction of Prof. PIERGIORGIO ODIFREDDI (Università di Torino) at Villa «La Querceta», Montecatini Terme (Pistoia), Italy, ***from June 20 to June 28, 1988.***

Courses

a) ***Overview of Computational Complexity Theory.*** (6 lectures in English).
Prof. Juris HARTMANIS (Cornell University, Ithaca).

Outline

Computational complexity theory is the study of the quantitative laws that govern computing. During the last twenty-five years, complexity theory has grown into a rich mathematical theory and today, it is one of the most active research areas in computer science. Among the most challenging open problems in complexity theory is the problem of understanding what is and is not feasible computable and more generally, a thorough understanding of the structure of the feasible computations. The best known of these problems is the classic P = ? NP problem. It is interesting to note that these problems, which were formulated in computer science, are actually basic problems about fundamental quantitative nature of mathematics. In essence, the P = ? NP problem is a question of how much harder is it to derive (computationally) a proof of a theorem than to check the validity of a proof.

The lectures on Computational Complexity will rewiew the basic concepts and techniques of complexity, summarize the earlier results and then review the more recent results about the structure of feasible computations.

b) ***Non-Traditional Logics for Computation.*** (6 lectures in English).
Prof. Anil NERODE (Cornell University, Ithaca).

Outline

A primer of non-traditional logics for non-experts. Many non-traditional logics are receiving wide attention in computer science because of potential importance in specification, development, and verification of programs and systems.

Prerequisites: some knowledge of undergraduate algebra, topology, computer science, and classical predicate logic. Otherwise self-contained.

I. Classical propositional and first order logic. Its models and Herbrand universes. Proof procedures of Gentzen natural deduction, Beth-Smullyan tableaux and resolution and unification. Discussion of automated classical propositional and first order reasoning.

II. Classical propositional and first order intuitionist logic. Its models such as Kripke models, Beth models, continuous function models, cpo models, Heyting valued models, categorical models. Models with discrete equality for Kroneckerian constructive algebra versus models with apartness for Brouwer's intuitionist analysis and Bishop's constructive analysis. Proof procedures of Gentzen, Fitting tableaux, natural deduction, resolution and unification for intuitionist logics. Discussion of automated intuitionist propositional and first order reasoning. Kleene's realizability for intuitionistic arithmetic as the archetype for lambda calculus realizability and the extraction of programs from proofs.

III. Modal first order logics, classical and intuitionist. The correspondence theory. Modal logics arising in computing such as algorithmic and dynamic logic, and temporal logics for description of concurrent processes and programs.

IV. Many sorted first order logics of all of the above sorts, model and proof theory and automation. Their use in algebraic specification theory, description of communicating processes, etc.

V. Logics for computing based on finite models, such as Gurevich's models for Pascal.

VI. Buchi's monadic second order theory of one successor and Rabin's second order monadic theory of two successors as languages for computing.

VII. Higher order intuitionist logic, de Bruijn and AUTOMATH, Constable and NUPRL, etc. Relation to rewrite rules and lambda calculus via the Curry-Howard isomorphism.

VIII. Other logics, such as logics of knowledge, probabilistic logics and uncertain reasoning logics.

c) ***Program verification.*** (6 lectures in English).
Prof. Richard PLATEK (Odissey Research Association, Ithaca).

Outline

In this lecture series we will review some of the fundamental logical theorems which form the basis of program verification. We will consider proofs of both sequential and concurrent programs. Some of the topics to be considered are:

I. Flowchart Verification: Floyd's verification condition theorem; its relationship to second order logic, infinitary logic, and PROLOG. Partial and total correctness; weakest precondition, strongest postcondition liberal and strict. Extensions of the method of invariants to include concurrency.

II. Hoare Logic: The verification of structured programs. Rules for constructs such as procedure call, recursion, pointers, loop exit statements, etc. Relative completeness results; incompleteness results. Extension of ordinary logic to a logic of partial terms in order to deal with undefined expressions (reading an uninitialized variable, indexing an array out of bounds, etc.).

III. Structured specification languages: An examination of Anna, a specification/assertion language for Ada.

IV. Concurrency: The Owicki-Gries approach; Hoare logic for CSP; the use of temporal logic.

There will also be a review of existing automated program verification systems.

d) ***Logic and Computer Science.*** (6 lectures in English).
Prof. Gerald SACKS (Harvard University, Cambridge, Mass).

Outline

- Logical foundations of prolog.
- Backtracking, cuts and operators.
- Prolog procedures.
- Database manipulation.
- Definite clause grammars and parsing.
- Classical recursion theory.

Basic references

- C. MARCUS, Prolog Programming, Addison-Wesley, 1986.
- I. BRATKO, Prolog Programming for Artificial Intelligence, Addison-Wesley, 1986.

e) ***Type Theory and Functional Programming.*** (6 lectures in English).
Prof. Andre SCEDROV (University of Pennsylvania, Philadelphia).

Outline

— Lambda Calculus (Typed, untyped, Church-Rosser Theorem, normal form; domains, fixed-points).
— Introduction to the ML language. (Syntax, typing rules, polymorphism, principal types, type-checking,...).
— Natural Deduction and the Proposition-as-Types Principle. (First-order minimal propositional calculus, normalization. second-order minimal propositional calculus, normalization, extensions to predicate calculi and arithmetic).
— Polymorphic Lambda Calculus. (Girard-Reynolds second-order lambda calculus, definable types: polyboole, polyint and other examples; expressiveness in terms of representable recursive functions, ...).
— Semantics. (Environment models, cartesian closed categories, realizability, coherent spaces, ...).
— Calculus of Constructions. (Syntax, typical ambiguity, examples, semantics, normalization).

Seminars

A number of seminars and special lectures will be offered during the Session.

FONDAZIONE C.I.M.E.
CENTRO INTERNAZIONALE MATEMATICO ESTIVO
INTERNATIONAL MATHEMATICAL SUMMER CENTER

"Global Geometry and Mathematical Physics"

is the subject of the Second 1988 C.I.M.E. Session.

The Session, sponsored by the Consiglio Nazionale delle Ricerche and the Ministero della Pubblica Istruzione, will take place under the scientific direction of Prof. MAURO FRANCAVIGLIA (Università di Torino), and Prof. FRANCESCO GHERARDELLI (Università di Firenze) at Villa «La Querceta», Montecatini (Pistoia), Italy, ***from July 4 to July 12, 1988.***

Courses

a) ***String Theory and Riemann Surfaces.*** (6 lectures in English).
Prof. L. ALVAREZ-GAUME (University of Boston, USA).

Contents

In these lectures we will review the new developments of string theory and its connections with the theory of Riemann surfaces, super-Riemann surfaces and their moduli spaces. The point of view taken will be to start with the conformal field theory formulation of string theory and then develop in detail the operator formalism for strings on higher genus surfaces. A tentative plane of the lectures is:

Lecture 1 - Introduction to string theory and conformal field theory (Part 1)
Lecture 2 - Introduction to string theory and conformal field theory (Part 2)
Lecture 3 - Perturbation theory for Bosonic strings. Belavin-Knizhnik theorem, Mumford forms, string infinities and the boundary of moduli spaces.
Lecture 4 - The operator formulation of string theory (Part 1)
Lecture 5 - The operator formulation of string theory (Part 2)
Lecture 6 - Virasoro action on moduli space, more general conformal theories, nonpertubative ideas and recent developments.

b) ***Riemann Surfaces and Infinite Grassmannians.*** (6 lectures in English).
Prof. E. ARBARELLO (Università di Roma "La Sapienza", Roma, Italy).

Contents

— Compact Riemann surfaces and their moduli (stable and semi-stable curves). Picard's group of the moduli space Mg. The Riemann-Roch-Grothendieck theorem.
— Kodaira-Spencer theory. Schiffer variations. Calculus of cohomology à la Mayer-Vietoris. The cotangent bundle of moduli space.
— Lie algebras *d* (Virasoro) and D (Virasoro-Heisenberg) and their relations with moduli space Mg.
— The Lie algebra gl_∞ and the "Boson-Fermion correspondence". Relations with the Lie algebras *d* and D. Calculation of the relevant cohomology.
— The infinite Grassmannian Gr and its geometry. The central extension of GL_∞. Tautological and determinant bundles; the function τ. Relation with Plücker's coordinates. The K.P. hierarchy.
— Krichever application. Relations between the functions τ and θ. The correlation function. The trisecting formula. Novikov conjecture.
— The sheaf of differential operators of order less or equal to one, acting on sections of the determinant bundle over Gr. Its restriction to moduli space.
— Some known results about moduli spaces of Riemann surfaces.

References

- L. ALVAREZ-GAUME, C. GOMEZ, C. REINA, Loop Groups, Grassmannians and String Theory, Phys. Lett. 190B; 55-62 (1987).
- E. ARBARELLO, M. CORNALBA, The Picard Groups of the Moduli Spaces of Curves, Topology 26(2), 153-171 (1987).
- E. ARBARELLO, C. DE CONCINI, V. KAC, C. PROCESI, Moduli Space of Curves and Representation Theory, preprint, 1987.
- A.A. BEILINSON, YU. I. MANIN, The Mumford Form and the Polyakov Measure in String Theory, Comm. Math. Phys. 107, 359-376 (1986).
- A.A. BEILINSON, YU. I. MANIN, V.V. SCHECHTMAN, Sheaves of the Virasoro and Neveu Schwarz Algebras, Moscow Univ. preprint, 1987.
- E. DATE, M. JIMBO, M. KASHIWARA, T. MIWA, Transformation Groups for Soliton Equations, in "Nonlinear Integrable Systems. Classical Theory and Quantum Theory", World Sci. (Singapore, 1983), pp. 39-119.
- J. HARER, The Second Homology Group of the Mapping Class Group of an Orientable Surface, Inv. Math. 72, 221-239 (1983).
- J. HARER, Stability of the Homology of the Mapping Class Group of Orientable Surfaces, Ann. of Math. 121, 215-249 (1985).
- V.G. KAC, D.H. PETERSON, Spin and Wedge Representations of Infinite Dimensional Lie Algebras and Groups, Proc. Nat. Acad. Sci. U.S.A. 78, 3308-3312 (1981).
- V.G. KAC, Highest Weight Representations of Conformal Current Algebras, in "Geometrical Methods in Field Theory", World Sci. (Singapore, 1986), pp. 3-15.
- N. KAWAMOTO, Y. NAMIKAWA, A. TSUCHIYA, Y. YAMADA, Geometric Realization of Conformal Field Theory on Riemann Surfaces, Nagoya Univ. preprint, 1987.
- YU. I. MANIN, Quantum String Theory and Algebraic Curves, Berkeley I.C.M. talk, 1986.
- E.Y. MILLER, The Homology of the Mapping Class Group, Journ. Diff. Geom. 24, 1-14 (1986).
- D. MUMFORD, Stability of Projective Varieties, L'Enseignement Mathém. 23, 39-110 (1977).
- A. PRESSLEY, G. SEGAL, Loop Groups, Oxford Univ. Press (Oxford, 1986).
- G. SEGAL, G. WILSON, Loop Groups and Equations of KdV Type, Publ. Math. I.H.E.S. 61, 3-64 (1985).
- C. VAFA, Conformal Theories and Punctured Surfaces, preprint, 1987.
- E. WITTEN, Quantum Field Theory, Grassmannians and Algebraic Curves, preprint, 1987.

c) ***The Topology and Geometry of Moduli Spaces.*** (6 lectures in English).
Prof. N.J. HITCHIN (Oxford University, Oxford, UK).

Contents

Topics will include:

(i) Moduli space instantons
(ii) Moduli space of monopoles
(iii) Moduli space of vortices
(iv) Teichmüller space
(v) Moduli spaces related to Riemann surfaces

References

- M.F. ATIYAH, Instantons in 2 and 4 Dimensions, Comm. Math. Phys. 93, 437-451 (1984).
- D. FREED, K. UHLENBECK, Instantons and 4 - Manifolds, Springer Verlag (Berlin, 1984).
- N.J. HITCHIN, A. KALBADE, U. LINDSTROM, M. ROCEK, Hyperkähler Metrics and Supersymmetry, Comm. Math. Phys. 108, 535-589 (1987).
- M.F. ATIYAH, N.J. HITCHIN, Geometry and Dynamics of Magnetic Monopoles, Princeton University Press (Princeton, 1988).
- A. JAPPE, C. TOMBES, Vortices and Monopoles, Birkhäuser (1980).
- N.J. HITCHIN, The Self-Duality Equation on a Riemann Surface, Proc. London Math. Soc. 55, 59-126 (1987).
- M.F. ATIYAH, R. BOTT, The Yang-Mills Equations over Riemann Surfaces, Phil. Trans. Roy. Soc. London, sec. A. 308, 523-615 (1982).

d) ***Differential Algebras in Field Theory.*** (6 lectures in English).
Prof. R. STORA (LAPP, Annecy-le-Vieux, France).

Contents

Lecture 1 - Introduction. The role of locality in perturbative quantum field theory ([1]).

Lecture 2 - The description of continuous symmetries in perturbative quantum field theories: global symmetries and their current algebras ([2]).

Lecture 3 - Current Algebra anomalies: algebraic structure. Local cohomologies of gauge Lie algebras ([3] - [10]).

Lecture 4 - Perturbative quantization of gauge theories, Slavnov-symmetries and the corresponding differential algebras ([6]).

Lecture 5 - More general differential algebras involving diffeomorphisms. Gravitational anomalies ([6], [7], [10]). Application to the first quantized string ([11]).

Lecture 6 - Miscellaneous examples:
(i) "Higher cocycles" in field theory: BRST current algebra versus Schwinger term (gauge Lie algebras extensions) ([12], [13]).
(ii) The "Torino" differential algebras for gravity and supergravity (tentative) ([14]).

References

The brief bibliography contains mainly review articles, from which the overwhelmingly voluminous original literature can be traced back.

[1] H. EPSTEIN, V. GLASER, Ann. Inst. H. Poincaré 19, 211 (1973).

[2] C. BECCHI, A. ROUET, R. STORA, Renormalizable Theories with Symmetry Breaking in "Field Theory Quantization and Statistical Physics", E. Tirapegui ed.; Reidel (Dordrecht), 1981).

[3] B. ZUMINO, in "Relativity, Groups and Topology. II", Les Houches XLIV, 1983; B. De Witt, R. Stora eds.; North-Holland (Amsterdam, 1984).

[4] R. STORA, in "Progress in Gauge Field Theory", Cargése 1983; G. 't Hooft et al. eds.; NATO ASI Series B 115, Plenum (New York, 1984).

[5] J. MANES, R. STORA, B. ZUMINO, Comm. Math, Phys. 102, 157 (1985).

[6] R. STORA, in "New Perspectives in Quantum Field Theories", XVIth GIFT Seminar, 1985; J. Abad et al. eds.; World Sci. (Singapore, 1986).

[7] L. ALVAREZ-GAUME, P. GINSPARG, Ann. Phys. 161, 423 (1985).

[8] M. DUBOIS-VIOLETTE, M. TALON, C.M. VIALLET, Comm. Math. Phys. 102, 105 (1985).

[9] L. BONORA, P. COTTA RAMUSINO, M. RINALDI, J. STASHEFF, CERN-Th 4647/87 and 4750/87 (to appear in Comm. Math. Phys.).

[10] L. ALVAREZ-GAUME, An Introduction to Anomalies, in "Erice 1985", G. Velo, A.S. Wightman eds.; ASI Series B 141, Plenum (New York, 1986).

[11] C. BECCHI, preprint, Genova 1987;
L. BAULIEU, M. BELLON, LPTHE 87-39.

[12] B. ZUMINO, Nucl. Phys. B 253, 477 (1985).

[13] L. BAULIEU, B. GROSSMANN, R. STORA, Phys. Lett. 180B, 95 (1986).

[14] T. REGGE, in "Relativity, Groups and Topology. II", Les Houches XLIV, 1983;
B. DE WITT, R. STORA eds.; North-Holland (Amsterdam, 1984).

Seminars

A number of seminars and special lectures will be offered during the Session.

LIST OF C.I.M.E. SEMINARS

		Publisher
1954 -	1. Analisi funzionale	C.I.M.E.
	2. Quadratura delle superficie e questioni connesse	"
	3. Equazioni differenziali non lineari	"
1955 -	4. Teorema di Riemann-Roch e questioni connesse	"
	5. Teoria dei numeri	"
	6. Topologia	"
	7. Teorie non linearizzate in elasticità, idrodinamica, aerodinamica	"
	8. Geometria proiettivo-differenziale	"
1956 -	9. Equazioni alle derivate parziali a caratteristiche reali	"
	10. Propagazione delle onde elettromagnetiche	"
	11. Teoria della funzioni di più variabili complesse e delle funzioni automorfe	"
1957 -	12. Geometria aritmetica e algebrica (2 vol.)	"
	13. Integrali singolari e questioni connesse	"
	14. Teoria della turbolenza (2 vol.)	"
1958 -	15. Vedute e problemi attuali in relatività generale	"
	16. Problemi di geometria differenziale in grande	"
	17. Il principio di minimo e le sue applicazioni alle equazioni funzionali	"
1959 -	18. Induzione e statistica	"
	19. Teoria algebrica dei meccanismi automatici (2 vol.)	"
	20. Gruppi, anelli di Lie e teoria della coomologia	"
1960 -	21. Sistemi dinamici e teoremi ergodici	"
	22. Forme differenziali e loro integrali	"
1961 -	23. Geometria del calcolo delle variazioni (2 vol.)	"
	24. Teoria delle distribuzioni	"
	25. Onde superficiali	"
1962 -	26. Topologia differenziale	"
	27. Autovalori e autosoluzioni	"
	28. Magnetofluidodinamica	"
1963 -	29. Equazioni differenziali astratte	"
	30. Funzioni e varietà complesse	"
	31. Proprietà di media e teoremi di confronto in Fisica Matematica	"

1964 -	32. Relatività generale	C.I.M.E.
	33. Dinamica dei gas rarefatti	"
	34. Alcune questioni di analisi numerica	"
	35. Equazioni differenziali non lineari	"
1965 -	36. Non-linear continuum theories	"
	37. Some aspects of ring theory	"
	38. Mathematical optimization in economics	"
1966 -	39. Calculus of variations	Ed. Cremonese, Firenz
	40. Economia matematica	"
	41. Classi caratteristiche e questioni connesse	"
	42. Some aspects of diffusion theory	"
1967 -	43. Modern questions of celestial mechanics	"
	44. Numerical analysis of partial differential equations	"
	45. Geometry of homogeneous bounded domains	"
1968 -	46. Controllability and observability	"
	47. Pseudo-differential operators	"
	48. Aspects of mathematical logic	"
1969 -	49. Potential theory	"
	50. Non-linear continuum theories in mechanics and physics and their applications	"
	51. Questions of algebraic varieties	"
1970 -	52. Relativistic fluid dynamics	"
	53. Theory of group representations and Fourier analysis	"
	54. Functional equations and inequalities	"
	55. Problems in non-linear analysis	"
1971 -	56. Stereodynamics	"
	57. Constructive aspects of functional analysis (2 vol.)	"
	58. Categories and commutative algebra	"
1972 -	59. Non-linear mechanics	"
	60. Finite geometric structures and their applications	"
	61. Geometric measure theory and minimal surfaces	"
1973 -	62. Complex analysis	"
	63. New variational techniques in mathematical physics	"
	64. Spectral analysis	"

1974 -	65. Stability problems	Ed. Cremonese, Firenze
	66. Singularities of analytic spaces	"
	67. Eigenvalues of non linear problems	"
1975 -	68. Theoretical computer sciences	"
	69. Model theory and applications	"
	70. Differential operators and manifolds	"
1976 -	71. Statistical Mechanics	Ed. Liguori, Napoli
	72. Hyperbolicity	"
	73. Differential topology	"
1977 -	74. Materials with memory	"
	75. Pseudodifferential operators with applications	"
	76. Algebraic surfaces	"
1978 -	77. Stochastic differential equations	"
	78. Dynamical systems	Ed. Liguori, Napoli and Birkhäuser Verlag
1979 -	79. Recursion theory and computational complexity	Ed. Liguori, Napoli
	80. Mathematics of biology	"
1980 -	81. Wave propagation	"
	82. Harmonic analysis and group representations	"
	83. Matroid theory and its applications	"
1981 -	84. Kinetic Theories and the Boltzmann Equation	(LNM 1048) Springer-Verlag
	85. Algebraic Threefolds	(LNM 947) "
	86. Nonlinear Filtering and Stochastic Control	(LNM 972) "
1982 -	87. Invariant Theory	(LNM 996) "
	88. Thermodynamics and Constitutive Equations	(LN Physics 228) "
	89. Fluid Dynamics	(LNM 1047) "
1983 -	90. Complete Intersections	(LNM 1092) "
	91. Bifurcation Theory and Applications	(LNM 1057) "
	92. Numerical Methods in Fluid Dynamics	(LNM 1127) "
1984	93. Harmonic Mappings and Minimal Immersions	(LNM 1161) "
	94. Schrödinger Operators	(LNM 1159) "
	95. Buildings and the Geometry of Diagrams	(LNM 1181) "
1985 -	96. Probability and Analysis	(LNM 1206) "
	97. Some Problems in Nonlinear Diffusion	(LNM 1224) "
	98. Theory of Moduli	(LNM 1337) "

<u>Note:</u> Volumes 1 to 38 are out of print. A few copies of volumes 23,28,31,32,33,34,36,38 are available on request from C.I.M.E.

1986 -	99. Inverse Problems	(LNM 1225)	Springer-Verlag
	100. Mathematical Economics	(LNM 1330)	"
	101. Combinatorial Optimization	to appear	"
1987 -	102. Relativistic Fluid Dynamics	to appear	"
	103. Topics in Calculus of Variations	(LNM 1365)	"
1988 -	104. Logic and Computer Science	to appear	"
	105. Global Geometry and Mathematical Physics	to appear	"

Vol. 1201: Curvature and Topology of Riemannian Manifolds. Proceedings, 1985. Edited by K. Shiohama, T. Sakai and T. Sunada. VII, 336 pages. 1986.

Vol. 1202: A. Dür, Möbius Functions, Incidence Algebras and Power Series Representations. XI, 134 pages. 1986.

Vol. 1203: Stochastic Processes and Their Applications. Proceedings, 1985. Edited by K. Itô and T. Hida. VI, 222 pages. 1986.

Vol. 1204: Séminaire de Probabilités XX, 1984/85. Proceedings. Edité par J. Azéma et M. Yor. V, 639 pages. 1986.

Vol. 1205: B.Z. Moroz, Analytic Arithmetic in Algebraic Number Fields. VII, 177 pages. 1986.

Vol. 1206: Probability and Analysis, Varenna (Como) 1985. Seminar. Edited by G. Letta and M. Pratelli. VIII, 280 pages. 1986.

Vol. 1207: P.H. Bérard, Spectral Geometry: Direct and Inverse Problems. With an Appendix by G. Besson. XIII, 272 pages. 1986.

Vol. 1208: S. Kaijser, J.W. Pelletier, Interpolation Functors and Duality. IV, 167 pages. 1986.

Vol. 1209: Differential Geometry, Peñíscola 1985. Proceedings. Edited by A.M. Naveira, A. Ferrández and F. Mascaró. VIII, 306 pages. 1986.

Vol. 1210: Probability Measures on Groups VIII. Proceedings, 1985. Edited by H. Heyer. X, 386 pages. 1986.

Vol. 1211: M.B. Sevryuk, Reversible Systems. V, 319 pages. 1986.

Vol. 1212: Stochastic Spatial Processes. Proceedings, 1984. Edited by P. Tautu. VIII, 311 pages. 1986.

Vol. 1213: L.G. Lewis, Jr., J.P. May, M. Steinberger, Equivariant Stable Homotopy Theory. IX, 538 pages. 1986.

Vol. 1214: Global Analysis – Studies and Applications II. Edited by Yu.G. Borisovich and Yu.E. Gliklikh. V, 275 pages. 1986.

Vol. 1215: Lectures in Probability and Statistics. Edited by G. del Pino and R. Rebolledo. V, 491 pages. 1986.

Vol. 1216: J. Kogan, Bifurcation of Extremals in Optimal Control. VIII, 106 pages. 1986.

Vol. 1217: Transformation Groups. Proceedings, 1985. Edited by S. Jackowski and K. Pawalowski. X, 396 pages. 1986.

Vol. 1218: Schrödinger Operators, Aarhus 1985. Seminar. Edited by E. Balslev. V, 222 pages. 1986.

Vol. 1219: R. Weissauer, Stabile Modulformen und Eisensteinreihen. III, 147 Seiten. 1986.

Vol. 1220: Séminaire d'Algèbre Paul Dubreil et Marie-Paule Malliavin. Proceedings, 1985. Edité par M.-P. Malliavin. IV, 200 pages. 1986.

Vol. 1221: Probability and Banach Spaces. Proceedings, 1985. Edited by J. Bastero and M. San Miguel. XI, 222 pages. 1986.

Vol. 1222: A. Katok, J.-M. Strelcyn, with the collaboration of F. Ledrappier and F. Przytycki, Invariant Manifolds, Entropy and Billiards; Smooth Maps with Singularities. VIII, 283 pages. 1986.

Vol. 1223: Differential Equations in Banach Spaces. Proceedings, 1985. Edited by A. Favini and E. Obrecht. VIII, 299 pages. 1986.

Vol. 1224: Nonlinear Diffusion Problems, Montecatini Terme 1985. Seminar. Edited by A. Fasano and M. Primicerio. VIII, 188 pages. 1986.

Vol. 1225: Inverse Problems, Montecatini Terme 1986. Seminar. Edited by G. Talenti. VIII, 204 pages. 1986.

Vol. 1226: A. Buium, Differential Function Fields and Moduli of Algebraic Varieties. IX, 146 pages. 1986.

Vol. 1227: H. Helson, The Spectral Theorem. VI, 104 pages. 1986.

Vol. 1228: Multigrid Methods II. Proceedings, 1985. Edited by W. Hackbusch and U. Trottenberg. VI, 336 pages. 1986.

Vol. 1229: O. Bratteli, Derivations, Dissipations and Group Actions on C*-algebras. IV, 277 pages. 1986.

Vol. 1230: Numerical Analysis. Proceedings, 1984. Edited by J.-P. Hennart. X, 234 pages. 1986.

Vol. 1231: E.-U. Gekeler, Drinfeld Modular Curves. XIV, 107 pages. 1986.

Vol. 1232: P.C. Schuur, Asymptotic Analysis of Soliton Problems. VIII, 180 pages. 1986.

Vol. 1233: Stability Problems for Stochastic Models. Proceedings, 1985. Edited by V.V. Kalashnikov, B. Penkov and V.M. Zolotarev. VI, 223 pages. 1986.

Vol. 1234: Combinatoire énumérative. Proceedings, 1985. Edité par G. Labelle et P. Leroux. XIV, 387 pages. 1986.

Vol. 1235: Séminaire de Théorie du Potentiel, Paris, No. 8. Directeurs: M. Brelot, G. Choquet et J. Deny. Rédacteurs: F. Hirsch et G. Mokobodzki. III, 209 pages. 1987.

Vol. 1236: Stochastic Partial Differential Equations and Applications. Proceedings, 1985. Edited by G. Da Prato and L. Tubaro. V, 257 pages. 1987.

Vol. 1237: Rational Approximation and its Applications in Mathematics and Physics. Proceedings, 1985. Edited by J. Gilewicz, M. Pindor and W. Siemaszko. XII, 350 pages. 1987.

Vol. 1238: M. Holz, K.-P. Podewski and K. Steffens, Injective Choice Functions. VI, 183 pages. 1987.

Vol. 1239: P. Vojta, Diophantine Approximations and Value Distribution Theory. X, 132 pages. 1987.

Vol. 1240: Number Theory, New York 1984–85. Seminar. Edited by D.V. Chudnovsky, G.V. Chudnovsky, H. Cohn and M.B. Nathanson. V, 324 pages. 1987.

Vol. 1241: L. Gårding, Singularities in Linear Wave Propagation. III, 125 pages. 1987.

Vol. 1242: Functional Analysis II, with Contributions by J. Hoffmann-Jørgensen et al. Edited by S. Kurepa, H. Kraljević and D. Butković. VII, 432 pages. 1987.

Vol. 1243: Non Commutative Harmonic Analysis and Lie Groups. Proceedings, 1985. Edited by J. Carmona, P. Delorme and M. Vergne. V, 309 pages. 1987.

Vol. 1244: W. Müller, Manifolds with Cusps of Rank One. XI, 158 pages. 1987.

Vol. 1245: S. Rallis, L-Functions and the Oscillator Representation. XVI, 239 pages. 1987.

Vol. 1246: Hodge Theory. Proceedings, 1985. Edited by E. Cattani, F. Guillén, A. Kaplan and F. Puerta. VII, 175 pages. 1987.

Vol. 1247: Séminaire de Probabilités XXI. Proceedings. Edité par J. Azéma, P.A. Meyer et M. Yor. IV, 579 pages. 1987.

Vol. 1248: Nonlinear Semigroups, Partial Differential Equations and Attractors. Proceedings, 1985. Edited by T.L. Gill and W.W. Zachary. IX, 185 pages. 1987.

Vol. 1249: I. van den Berg, Nonstandard Asymptotic Analysis. IX, 187 pages. 1987.

Vol. 1250: Stochastic Processes – Mathematics and Physics II. Proceedings 1985. Edited by S. Albeverio, Ph. Blanchard and L. Streit. VI, 359 pages. 1987.

Vol. 1251: Differential Geometric Methods in Mathematical Physics. Proceedings, 1985. Edited by P.L. García and A. Pérez-Rendón. VII, 300 pages. 1987.

Vol. 1252: T. Kaise, Représentations de Weil et GL_2 Algèbres de division et GL_n. VII, 203 pages. 1987.

Vol. 1253: J. Fischer, An Approach to the Selberg Trace Formula via the Selberg Zeta-Function. III, 184 pages. 1987.

Vol. 1254: S. Gelbart, I. Piatetski-Shapiro, S. Rallis. Explicit Constructions of Automorphic L-Functions. VI, 152 pages. 1987.

Vol. 1255: Differential Geometry and Differential Equations. Proceedings, 1985. Edited by C. Gu, M. Berger and R.L. Bryant. XII, 243 pages. 1987.

Vol. 1256: Pseudo-Differential Operators. Proceedings, 1986. Edited by H.O. Cordes, B. Gramsch and H. Widom. X, 479 pages. 1987.

Vol. 1257: X. Wang, On the C*-Algebras of Foliations in the Plane. V, 165 pages. 1987.

Vol. 1258: J. Weidmann, Spectral Theory of Ordinary Differential Operators. VI, 303 pages. 1987.

Vol. 1259: F. Cano Torres, Desingularization Strategies for Three-Dimensional Vector Fields. IX, 189 pages. 1987.

Vol. 1260: N.H. Pavel, Nonlinear Evolution Operators and Semi-groups. VI, 285 pages. 1987.

Vol. 1261: H. Abels, Finite Presentability of S-Arithmetic Groups. Compact Presentability of Solvable Groups. VI, 178 pages. 1987.

Vol. 1262: E. Hlawka (Hrsg.), Zahlentheoretische Analysis II. Seminar, 1984–86. V, 158 Seiten. 1987.

Vol. 1263: V.L. Hansen (Ed.), Differential Geometry. Proceedings, 1985. XI, 288 pages. 1987.

Vol. 1264: Wu Wen-tsün, Rational Homotopy Type. VIII, 219 pages. 1987.

Vol. 1265: W. Van Assche, Asymptotics for Orthogonal Polynomials. VI, 201 pages. 1987.

Vol. 1266: F. Ghione, C. Peskine, E. Sernesi (Eds.), Space Curves. Proceedings, 1985. VI, 272 pages. 1987.

Vol. 1267: J. Lindenstrauss, V.D. Milman (Eds.), Geometrical Aspects of Functional Analysis. Seminar. VII, 212 pages. 1987.

Vol. 1268: S.G. Krantz (Ed.), Complex Analysis. Seminar, 1986. VII, 195 pages. 1987.

Vol. 1269: M. Shiota, Nash Manifolds. VI, 223 pages. 1987.

Vol. 1270: C. Carasso, P.-A. Raviart, D. Serre (Eds.), Nonlinear Hyperbolic Problems. Proceedings, 1986. XV, 341 pages. 1987.

Vol. 1271: A.M. Cohen, W.H. Hesselink, W.L.J. van der Kallen, J.R. Strooker (Eds.), Algebraic Groups Utrecht 1986. Proceedings. XII, 284 pages. 1987.

Vol. 1272: M.S. Livšic, L.L. Waksman, Commuting Nonselfadjoint Operators in Hilbert Space. III, 115 pages. 1987.

Vol. 1273: G.-M. Greuel, G. Trautmann (Eds.), Singularities, Representation of Algebras, and Vector Bundles. Proceedings, 1985. XIV, 383 pages. 1987.

Vol. 1274: N. C. Phillips, Equivariant K-Theory and Freeness of Group Actions on C*-Algebras. VIII, 371 pages. 1987.

Vol. 1275: C.A. Berenstein (Ed.), Complex Analysis I. Proceedings, 1985–86. XV, 331 pages. 1987.

Vol. 1276: C.A. Berenstein (Ed.), Complex Analysis II. Proceedings, 1985–86. IX, 320 pages. 1987.

Vol. 1277: C.A. Berenstein (Ed.), Complex Analysis III. Proceedings, 1985–86. X, 350 pages. 1987.

Vol. 1278: S.S. Koh (Ed.), Invariant Theory. Proceedings, 1985. V, 102 pages. 1987.

Vol. 1279: D. Ieşan, Saint-Venant's Problem. VIII, 162 Seiten. 1987.

Vol. 1280: E. Neher, Jordan Triple Systems by the Grid Approach. XII, 193 pages. 1987.

Vol. 1281: O.H. Kegel, F. Menegazzo, G. Zacher (Eds.), Group Theory. Proceedings, 1986. VII, 179 pages. 1987.

Vol. 1282: D.E. Handelman, Positive Polynomials, Convex Integral Polytopes, and a Random Walk Problem. XI, 136 pages. 1987.

Vol. 1283: S. Mardešić, J. Segal (Eds.), Geometric Topology and Shape Theory. Proceedings, 1986. V, 261 pages. 1987.

Vol. 1284: B.H. Matzat, Konstruktive Galoistheorie. X, 286 pages. 1987.

Vol. 1285: I.W. Knowles, Y. Saitō (Eds.), Differential Equations and Mathematical Physics. Proceedings, 1986. XVI, 499 pages. 1987.

Vol. 1286: H.R. Miller, D.C. Ravenel (Eds.), Algebraic Topology. Proceedings, 1986. VII, 341 pages. 1987.

Vol. 1287: E.B. Saff (Ed.), Approximation Theory, Tampa. Proceedings, 1985–1986. V, 228 pages. 1987.

Vol. 1288: Yu. L. Rodin, Generalized Analytic Functions on Riemann Surfaces. V, 128 pages, 1987.

Vol. 1289: Yu. I. Manin (Ed.), K-Theory, Arithmetic and Geometry. Seminar, 1984–1986. V, 399 pages. 1987.

Vol. 1290: G. Wüstholz (Ed.), Diophantine Approximation and Transcendence Theory. Seminar, 1985. V, 243 pages. 1987.

Vol. 1291: C. Mœglin, M.-F. Vignéras, J.-L. Waldspurger, Correspondances de Howe sur un Corps p-adique. VII, 163 pages. 1987

Vol. 1292: J.T. Baldwin (Ed.), Classification Theory. Proceedings, 1985. VI, 500 pages. 1987.

Vol. 1293: W. Ebeling, The Monodromy Groups of Isolated Singularities of Complete Intersections. XIV, 153 pages. 1987.

Vol. 1294: M. Queffélec, Substitution Dynamical Systems – Spectral Analysis. XIII, 240 pages. 1987.

Vol. 1295: P. Lelong, P. Dolbeault, H. Skoda (Réd.), Séminaire d'Analyse P. Lelong – P. Dolbeault – H. Skoda. Seminar, 1985/1986. VII, 283 pages. 1987.

Vol. 1296: M.-P. Malliavin (Ed.), Séminaire d'Algèbre Paul Dubreil et Marie-Paule Malliavin. Proceedings, 1986. IV, 324 pages. 1987.

Vol. 1297: Zhu Y.-l., Guo B.-y. (Eds.), Numerical Methods for Partial Differential Equations. Proceedings. XI, 244 pages. 1987.

Vol. 1298: J. Aguadé, R. Kane (Eds.), Algebraic Topology, Barcelona 1986. Proceedings. X, 255 pages. 1987.

Vol. 1299: S. Watanabe, Yu.V. Prokhorov (Eds.), Probability Theory and Mathematical Statistics. Proceedings, 1986. VIII, 589 pages. 1988.

Vol. 1300: G.B. Seligman, Constructions of Lie Algebras and their Modules. VI, 190 pages. 1988.

Vol. 1301: N. Schappacher, Periods of Hecke Characters. XV, 160 pages. 1988.

Vol. 1302: M. Cwikel, J. Peetre, Y. Sagher, H. Wallin (Eds.), Function Spaces and Applications. Proceedings, 1986. VI, 445 pages. 1988.

Vol. 1303: L. Accardi, W. von Waldenfels (Eds.), Quantum Probability and Applications III. Proceedings, 1987. VI, 373 pages. 1988.

Vol. 1304: F.Q. Gouvêa, Arithmetic of p-adic Modular Forms. VIII, 121 pages. 1988.

Vol. 1305: D.S. Lubinsky, E.B. Saff, Strong Asymptotics for Extremal Polynomials Associated with Weights on ℝ. VII, 153 pages. 1988.

Vol. 1306: S.S. Chern (Ed.), Partial Differential Equations. Proceedings, 1986. VI, 294 pages. 1988.

Vol. 1307: T. Murai, A Real Variable Method for the Cauchy Transform, and Analytic Capacity. VIII, 133 pages. 1988.

Vol. 1308: P. Imkeller, Two-Parameter Martingales and Their Quadratic Variation. IV, 177 pages. 1988.

Vol. 1309: B. Fiedler, Global Bifurcation of Periodic Solutions with Symmetry. VIII, 144 pages. 1988.

Vol. 1310: O.A. Laudal, G. Pfister, Local Moduli and Singularities. V, 117 pages. 1988.

Vol. 1311: A. Holme, R. Speiser (Eds.), Algebraic Geometry, Sundance 1986. Proceedings. VI, 320 pages. 1988.

Vol. 1312: N.A. Shirokov, Analytic Functions Smooth up to the Boundary. III, 213 pages. 1988.

Vol. 1313: F. Colonius, Optimal Periodic Control. VI, 177 pages. 1988.

Vol. 1314: A. Futaki, Kähler-Einstein Metrics and Integral Invariants. IV, 140 pages. 1988.

Vol. 1315: R.A. McCoy, I. Ntantu, Topological Properties of Spaces of Continuous Functions. IV, 124 pages. 1988.

Vol. 1316: H. Korezlioglu, A.S. Ustunel (Eds.), Stochastic Analysis and Related Topics. Proceedings, 1986. V, 371 pages. 1988.

Vol. 1317: J. Lindenstrauss, V.D. Milman (Eds.), Geometric Aspects of Functional Analysis. Seminar, 1986–87. VII, 289 pages. 1988.

Vol. 1318: Y. Felix (Ed.), Algebraic Topology – Rational Homotopy. Proceedings, 1986. VIII, 245 pages. 1988

Vol. 1319: M. Vuorinen, Conformal Geometry and Quasiregular Mappings. XIX, 209 pages. 1988.